A Guide to
Historic Coal Towns
of the Big Sandy
River Valley

A Guide to HISTORIC COAL TOWNS of the BIG SANDY RIVER VALLEY

George D. Torok

The University of Tennessee Press ■ Knoxville

Unless otherwise noted, the photographs were taken by the author.

This book is printed on acid-free paper.

Library of Congress Cataloging-in-Publication Data

Torok, George D.
  A guide to historic coal towns of the Big Sandy River Valley /
  George D. Torok.— 1st ed.
      p. cm.
Includes bibliographical references (p.) and index.
ISBN 1-57233-282-4 (pbk.)

1. Big Sandy River Valley (Ky. and W. Va.)—Guidebooks.
2. Historic sites—Big Sandy River Valley (Ky. and W. Va.)—Guidebooks.
3. Cities and towns—Big Sandy River Valley (Ky. and W. Va.)—Guidebooks.
4. Mining camps—Big Sandy River Valley (Ky. and W. Va.)—Guidebooks.
5. Big Sandy River Valley (Ky. and W. Va.)—History, Local.
6. Coal mines and mining—Big Sandy River Valley (Ky. and W. Va.)—History.
I. Title.

F457.B5T67 2004
917.69'15404—dc22   2003025095

To Blanca, who shared in all the wonders of the Big Sandy River Valley, I lovingly dedicate this book.

Contents

Illustrations

Preface

This guidebook presents an overview of the coal mining heritage of the Big Sandy River Valley and its many coal mining towns, including a general history of the valley and the great Appalachian coal boom that took place from the 1880s to the 1920s. A variety of primary sources, interviews, scholarly works, and local, regional, and industrial histories were used to create as complete a picture as possible. However, this volume is not intended to be a definitive history or an exhaustive guide to the entire region. The coal industry has undergone many changes, and areas discussed here have been operated sporadically under many different companies and proprietors. Companies that were the most important in developing these towns have been included. No doubt there are occasional errors and discrepancies, and these will encourage others to continue studying the region. Perhaps this guidebook will be a starting point for future studies, research, preservation, and site development.

Only a small fraction of the hundreds of coal towns that were once scattered throughout region remain today. Most no longer exist, and those that have survived have suffered great structural loss. This guidebook offers a glimpse of each site as it appeared during the late 1990s; many have changed significantly since then. Some structures were demolished as this work was being researched, written, and published.

Most of the sites in this guidebook lie along the major highways of the Big Sandy region: U.S. 119, U.S. 23, and U.S. 52. Some surrounding areas have also been included because of their close proximity, their importance in the development of the region, or the abundance of structures and sites that remain. These additional areas lie outside of the Big

Sandy Valley but can easily be visited when traveling along these routes. Exciting developments are under way throughout the region. Harlan County, Kentucky, west of the Big Sandy River Valley, played a major role in the development of coal mining, and the company towns of Benham and Lynch have recently undergone extensive renovation and restoration. They are in the process of becoming centers of heritage tourism, with the opening of museum, mining, and travel facilities. The Big Stone Gap coalfields, south of the valley on U.S. 23, also have several important coal towns and museums. The eleven-county section of West Virginia that has been declared a National Coal Heritage Area has opened that region to heritage tourism projects. Several sites in the Heritage Area along U.S. 52 south of the Big Sandy River Valley have also been included.

Most of the sites described in this work are accessible from state and county highways but are almost all located on private property. Visitors can observe them but are warned not to trespass or enter enclosed or posted areas. Even accessible areas can be extremely dangerous. In some communities that are being developed for tourism, there are mine sites, equipment, and structures that have been abandoned or have yet to be restored. In many coal towns, remaining structures may have weakened foundations, debris, or crumbling walls and supports. These buildings should not be entered unless they are specifically open to visitors. Under no conditions should a mine site be entered.

Many people assisted with the preparation of this guidebook. Dr. Thomas Matijasic, history professor at Prestonsburg Community College, helped with some of the initial research and writing. He laid out a basic outline of the history of the region and wrote the initial drafts for some of the entries on Johnson, Floyd, and Lawrence Counties. L. Martin "Marty" Perry, National Register site coordinator at the Kentucky Heritage Council, read through several early drafts and helped find funding for research. Some of the initial survey work was done under a 1992 University of Kentucky Teaching Improvement Grant. A 1996 Kentucky Heritage Council Grant, administered through the Big Sandy Area Development District offices in Prestonsburg, Kentucky, permitted more detailed surveys of many of the sites. A 1999 Kentucky Heritage Council Grant allowed for extended research and the photographing of sites. El Paso Community College's Division of Social Sciences, under the direction of Chairman Saul Candelas and Dean Susana Rodarte, helped me present research at several conferences and prepare journal articles. The Jenkins Heritage Commission and the Paintsville Tourism Commission made financial contributions and encouraged the project.

Several readers evaluated drafts of the manuscript and offered useful suggestions: Susan Braselton of Preservation Partners Inc., Arlington, Virginia; Dr. Michael E. Workman of the Institute for the History of Technology and Industrial Archeology at West Virginia University; Dr. Stuart McGehee of the Eastern Regional Coal Archives, Craft Memorial Library, Bluefield, West Virginia; and Dr. Carroll Van West of the Historic Preservation Program, Middle Tennessee State University. And of course Joyce Harrison, acquisitions editor at the University of Tennessee Press, who saw the potential value of this guidebook and encouraged its publication.

Many others provided great assistance and deserve special mention. The librarians and support staff of Prestonsburg Community College and El Paso Community College were especially helpful. Peggy Bentley of the Mary Jo Wolfe Memorial Library in Jenkins, Kentucky, was one of the first people to assist with the project and made many valuable and enthusiastic contributions. Billy Sue Davis of the Pikeville City Library provided many useful references, access to the Paul Mays Collection, and financial support. Others who offered assistance were Bobbie Gothard of the Kentucky Coal Mining Museum in Benham, Kentucky; Rhonda Simpson of Gulnare, Kentucky, who helped with photography and research; Sandra Looney of Seco, Kentucky, who helped provide information on the town and its development; John "Dobbin" Owens, who shared his memories of Pike County; Mayor Bobby Dorton and Louis Heneger of Appalachia, Virginia; David Zegeer, who provided useful information on Jenkins; David Reynolds of the Matewan Development Center; Tim Tingle of the Kentucky Department of Library and Archives in Frankfort; D. James Keenan of Stone, Kentucky; Katherine M. Jourdan, National Register coordinator, West Virginia Division of Culture and History; members of the Norfolk & Western Historical Society, Roanoke, Virginia; the Chesapeake & Ohio Historical Society, Clifton Forge, Virginia; and Lillian Martinez, periodicals librarian at the University of Texas at El Paso.

I extend a special thanks to all the people I met and visited with in the Big Sandy region who gave directions, provided guidance through sites, and shared their memories. Without them this work would not have been possible.

1   The Big Sandy River Valley

Introduction

The Big Sandy River Valley is located in southern Appalachia. It was a remote mountain area and one of the last regions to be settled in the United States. Isolated and lightly populated during most of the 1800s, the region underwent a great industrial transformation as the century drew to a close. Once railroads penetrated the mountain hollows of the Big Sandy River Valley, its rich seams of bituminous coal were finally within reach of coal operators and mining companies. The area attracted investors, speculators, capital, and corporations. Isolated mountain communities became the sites of great mining operations, small regional commercial centers boomed, and hundreds of coal company towns appeared almost overnight. These coal towns became vibrant, thriving communities where countless thousands of people came to live and work. Native white mountaineers soon found themselves surrounded by eastern developers, southern laborers, European immigrants, and African Americans. By the 1920s, the Big Sandy River Valley was a well-populated region connected by a web of railroads and new highways to the outside world. It played a critical role in the industrial development of the nation and prospered well into the mid-twentieth century. The people of the region mined and transported coal but also raised families, built communities, suffered through the Great Depression, fought in two world wars, and made valuable contributions to our national culture.

The great coal boom continued into the 1940s, but the period following the Second World War brought decline. By the 1960s, the region experienced hard times and became well known for its poverty and isolation once again.

Today, many of these once booming mining communities have faded away and are rapidly becoming ghost towns. Their populations are a fraction of what they once were, and buildings, homes, and mine sites have been abandoned. Many Big Sandy River Valley coal towns no longer exist. They have been mined out, bulldozed under, burned down, or taken over by the land.

Geography

The political boundaries of the Big Sandy River Valley spread over three states: Kentucky, West Virginia, and Virginia. The valley includes the Kentucky counties of Lawrence, Johnson, Martin, Floyd, Pike, and Letcher, and the West Virginia counties of Wayne, Mingo, and McDowell. It is approximately 190 miles long and 80 miles wide and comprises two major tributaries: the Tug and Levisa Forks (see map on pages 4 and 5). The headwaters of these two forks originate in southwest Virginia and southern West Virginia. The deeper and wider of the two is the Tug Fork, which becomes a major tributary in McDowell County, West Virginia, and forms the West Virginia–Kentucky border for 94 miles of its length. The Levisa Fork winds through the western part of the valley, running north from Pike County, Kentucky. It is fed by two major tributaries, Russell Fork and Elkhorn Creek. Russell Fork drains mountains south of Pike County, while Elkhorn Creek originates in the Kentucky-Virginia border region around Letcher County. The Levisa and Tug Forks come together at Louisa, Kentucky, forming the Big Sandy River, which continues north to meet the Ohio River at Catlettsburg. The Kentucky, Clinch, and Powell Rivers also pass through parts of the region. Although they do not feed into the Big Sandy, they have played a vital role in the development of the area. The Kentucky River is fed by three major forks and flows north to Frankfort. The Clinch and Powell Rivers lie beyond Pound Gap and run through Wise County, Virginia. Many of the Big Sandy region's coal towns lie along these rivers or the countless streams that feed them.[1]

The Big Sandy region lies within the Cumberland Mountains that extend from the western ridge of the Alleghenies into eastern Kentucky and southwest Virginia. Whitesburg, Kentucky, attorney and writer Harry Caudill described the area as a "serrated upland . . . its jagged hills and narrow winding valleys cover some ten thousand square miles."[2] In the early-twentieth-century John Fox Jr., author of the *Trail of the Lonesome*

Pine, a romantic account of the mountains south of Letcher County, Kentucky, described the range as one that "runs its mighty ribs into the Cumberland range with such humiliating violence that the Cumberland was turned feet over head by the shock" and "where there [are] rich veins of pure coking coal and not an ounce of iron ore; to the south is the iron ore and not an ounce of coal."[3] The mountains are part of a great westerly sloping plateau, formed about 350 to 300 million years ago and cut by streams over millions of years.

The Cumberland range hosts a great variety of altitudes and terrain. The elevations along the western edge of West Virginia are part of the Allegheny Plateau, whose waters feed the Tug Fork of the Big Sandy River. The valley has very uneven topography. The northern Big Sandy region has fewer deeply cut valleys and much more flatland, while in the south there are steep ravines and narrow valleys. Altitudes range from five hundred feet at Louisa, where the Tug and Levisa Forks meet, to mountain peaks over four thousand feet in southwest Virginia.

The great range of terrain in the region can be experienced in several scenic locations. In southwest Virginia, the Russell Fork of the Big Sandy has worn a deep river gorge. At Pine Mountain River Canyon, a five-mile stretch of the Russell Fork, the waters descend 350 feet over rocks and boulders. Rising 1,600 feet above the gorge is a mountain called The Towers, which has been a prominent landmark since prehistoric times. The path of the fork has made a series of rapids and small waterfalls known as the Breaks of Sandy, today part of the Breaks Interstate Park on the Virginia-Kentucky border, near U.S. 460.[4]

Another great place to view the Cumberland Range is at High Knob, in the Daniel Boone National Forest off U.S. 23 in Wise County, Virginia. An observation deck at the 4,162-foot summit allows breathtaking views of the surrounding mountains of southern Appalachia. On a crisp, clear autumn day, areas of four nearby states can be seen from this point. In Harlan County, Kentucky, at the eastern end of the Cumberland Mountains, a drive along KY 179 offers a similar view. The road crosses Big Black Mountain, rising 4,145 feet, allowing stunning views of east Kentucky and southwest Virginia. While the entire region has a wide range of altitudes, most of the coal towns in this guide are located along the valley floor, usually at altitudes between 800 and 1,400 feet.[5]

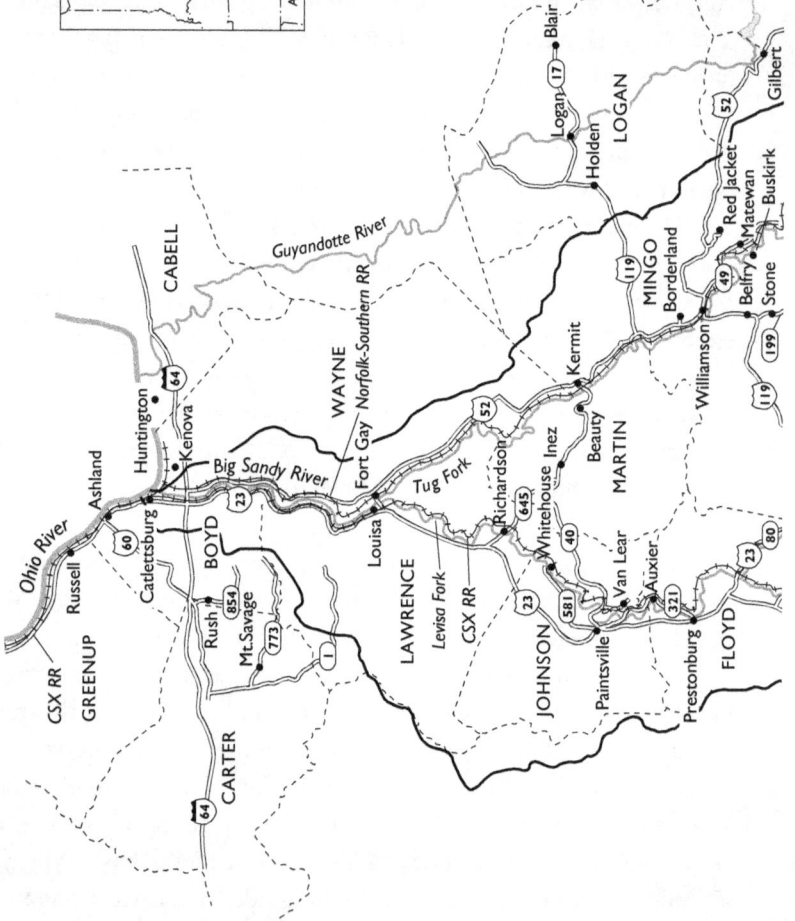

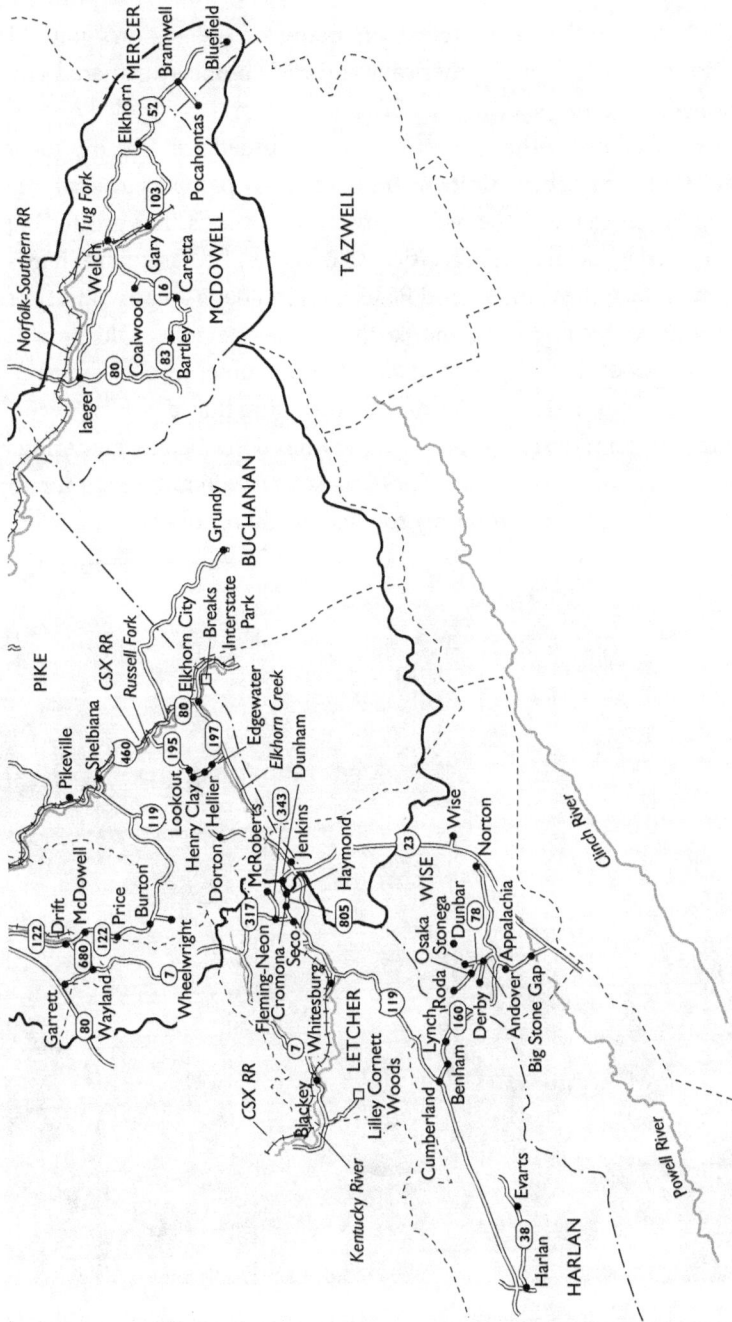

Big Sandy River Valley

Natural Resources

Big Sandy Valley settlers have relied on many of the region's natural resources for centuries. Native Americans and early pioneers gathered salt, iron ore, oil, tar, and saltpeter. Some of these resources were traded or sold as a source of income, others were simply consumed. One of the most abundant resources has been timber, and many early pioneers lived in traditional log cabins made of the region's many hardwoods. The greater Big Sandy River Valley has native vegetation that is classified as central hardwood forest. In fact, the Cumberland Plateau region has a greater variety of forest trees than any region of the earth's temperate zone. Old-growth hardwood forests of oak, poplar, maple, sycamore, pine, spruce, chestnut, and walnut trees covered the valley and surrounding hillsides. Evergreens, the huge cedars, hemlocks, and pines, sometimes described as mysterious and moody, were also common.[6] These trees were often giants, four to eight feet in diameter, and commonly reached heights of over one hundred

The Russell Fork of the Big Sandy River winds through Breaks Interstate Park on the Kentucky-Virginia border.

feet. It was said that some pioneer families lived in hollowed sycamores until they built their homes. Some trees were of spectacular proportions. The Mingo Oak, located near Williamson, West Virginia, was estimated to be 582 years old when it died in 1938. It was the largest white oak in the United States, with a height of 146 feet and a diameter of almost 10 feet. The tree was a local tourist attraction, and a West Virginia historical highway marker is now on the spot where this majestic tree once stood.[7] Unfortunately, few trees of such proportions remain today. Once densely forested, the Big Sandy region suffered greatly during the timber boom of the late nineteenth century. Most of these trees fell to the logging industry, although some reforestation efforts did produced new-growth forests. This timber was used to build barns, docks, commercial structures, and fences. When coal mining began in the valley, lumber was needed to build houses in company towns and support beams in the mines. The old-growth forests were quickly depleted. Today, the most common trees found are yellow poplar, spruce, sycamore, oak, chestnut, and walnut. The tulip tree, or yellow poplar, a member of the magnolia family, is probably the most widely used for lumber. In higher elevations, where the soils are derived from sandstone and shale, oak and chestnut types are common. Farther down the mountainside, on the lower slopes, white oak and pine are found in abundance. Very little virgin timberland exists today, but an example of an old-growth forest can be experienced in the Lilley Cornett Woods, south of Blackey in Letcher County, Kentucky, off KY 1103.[8]

Although the variety is limited, road cuts along the highways of the Big Sandy River Valley reveal thick layers of rock. The most common are layered sedimentary rocks (sandstone, shale, and limestone) that often contains coal seams. Sandstone can be found in several colors (brown, red, pink, or gray) and textures and is frequently filled with fossilized remains. Shales may contain fossilized plants, tree trunks, leaves, and coal. Limestone is less common and is usually found in the western part of the valley, especially along the banks of the tributaries of the Kentucky River. These stones have had many uses over the years. Quarried stones have long been used in construction and can be seen in foundations, chimneys, walls, and roads. Today, quarries and gravel pits continue to provide building materials. Although homes were rarely made exclusively of stone, commercial and coal company buildings throughout the region did use native stone, especially sandstone, in construction. The Big Sandy region also has natural gas and petroleum. Iron, common in the northern end of the valley near the Ohio River, led to the development of an iron industry in Ashland and Ironton in the mid-nineteenth century.

Of course, the most important natural resource in the Big Sandy region has been coal. The valley has coal seams averaging four to five feet thick lying in broad areas known as coalfields or coal beds. These coalfields were formed in the Pennsylvania period, the great coal age (320 to 280 million years ago). The Big Sandy region of West Virginia contains the Williamson coalfields and borders three other large fields, the Logan, the Kanawha, and the Flat Top-Pocahontas. In east Kentucky, the Elkhorn coalfield consists of several seams that dominate the Big Sandy Reserve District. This coal is heavily concentrated in Johnson, Floyd, and Pike Counties. Parts of Letcher and Harlan Counties lie within the Upper Cumberland and the Hazard Reserve Districts.[9] The coal found throughout the Big Sandy River Valley is generally high-quality bituminous coal.

The soils of the valley are generally thin and do not contain many minerals, partly because of the sandstone base that underlies much of the region. Poor soil, in combination with hilly terrain, has limited agriculture in the Big Sandy area. The only good land for growing crops lies on the valley floor alongside the rivers and creeks. These narrow bands of bottomland, usually a few hundred feet wide, provide good soils and are periodically replenished by flooding. The hillsides have not been easy to farm. They may have good soil, but they contain many rocks and are difficult to plow, plant, and harvest. Even where good soil is available, limited transportation and markets have restricted farming in the valley. Pioneers grew only what they needed to survive and seldom had a chance to make a profit. Although not particularly good for planting, some areas of the valley were once used for grazing. Livestock were more common before 1900, and there was some limited regional trade among farmers.[10] Today, only the northern portion of the river valley has any significant farming.

Flora and Fauna

The Big Sandy region has a wealth of plant life and some areas are quite lush. Along with a great variety of trees, there are many beautiful plants and flowers. Evergreens cap many of the mountain peaks and mosses, and ferns flourish on the forested hillsides and along rivers and streams. Flowers begin to appear in May, and by June there is great floral beauty in the mountains. The rhododendron, a flower of pink and white or rose-purple, is one of the most common in the region. Several varieties of magnolia produce white, sweet-scented blossoms. Yellow poplar yields dainty green chalices. The great laurel, or mountain rosebay, covers the hills and cliffs with bell-shaped rose-purple flowers. By late spring, the mountains bloom with bloodroot, bluebell, wild ginger, moss pink, dogwood, and dozens of other species. The mountain laurel, or calico bush, is at its loveliest in June, and during summer the blossoms of wild strawberry, papaw, wild grape, haw, and serviceberry mature. Chestnuts, hazelnuts, hickory nuts, walnuts, and beechnuts begin to appear by fall.[11]

Not all the vegetation on the Big Sandy landscape is native to the area. Some plants, trees, and flowers have been introduced over the years, especially during reforestation projects carried out in the 1930s under the Civilian Conservation Corp (CCC). One of the most prominent new forms of vegetation in southern Appalachia is kudzu, a fast-growing climbing vine introduced to Louisiana in the late nineteenth century. It was brought from Japan and was planted widely in the South in the 1930s as a way to control erosion. Kudzu's bright green leaves cover many mountainsides in spring and summer, but it is extremely difficult to contain or exterminate. While it has helped in some areas with erosion problems, it has grown out of control in many others. Visitors to the region will see hillsides blanketed with kudzu often burying homes, mines, and other structures.[12]

Unfortunately, native plant life has been greatly affected by mining in the twentieth century. As mountaintops were stripped and augered for coal, the landscape was flattened, removing native vegetation and changing local drainage patterns. In many cases, lands were simply abandoned after the coal was removed. Strip mining has left the mountains scarred, created mud slides and floods, and released toxic substances into the air and water supply. The 1977 Surface Mining Control and Reclamation Act (more commonly known as the Federal Strip Mining Act) now requires coal operators to reclaim lands. Stripped lands are contoured to simulate their original appearance and plant life and vegetation are restored. Although this has helped bring back some of the natural landscape of the region, mining coal through mountaintop removal, a method which has become more common in the late twentieth century, has created more serious problems. Many visitors to the region are unaware of strip mining and mountaintop operations unless they examine the landscape from the air or look down from some of the higher elevations.

Poor enforcement of environmental laws has poisoned streams and introduced toxins into the water table. As in any area which undergoes industrial development, the number and variety of animal and plant life has noticeably declined. The American bison, which once roamed the area, was driven into Ohio by the mid-nineteenth century. Gray timber wolves, panthers, and elk, once common in the valley, entirely disappeared during the twentieth century. Today, deer and occasional black bear are the only large animals seen in the region. Small game are far more numerous, especially raccoon, fox, otter, opossum, red and gray squirrels, rabbits, and woodchuck. The many streams and rivers offer a variety of fish. The most common are the large and smallmouth bass, channel catfish, trout, and crappie.

Climate and Weather

Travelers can enjoy the Big Sandy region year round. The climate is generally temperate and mild. There is a long growing season lasting almost six months. Precipitation is plentiful and sometimes excessive, causing floods in many areas. The first frost arrives in early October, and freezing temperatures are rare by late April. At times, temperatures can be extreme, but seldom for extended periods. In the summer months, daytime temperatures rarely reach one hundred degrees. In winter months the temperature seldom dips below zero, and it rises significantly during the daylight hours.

For visitors to the Big Sandy River region, each season has its own advantages and disadvantages. Spring is a beautiful time to visit, with flowers in bloom by early May and warm daytime temperatures. While pleasant, May and June can also bring sudden violent thunderstorms and the potential for flash floods. Summer months have warmer temperatures and periods of humidity that can sometimes be quite oppressive, but foliage and vegetation are in full bloom. Fall is probably the best time to visit. Temperatures are cooler and the humidity is much lower. The leaves turn color after the first burst of cold Canadian air in mid-October. Because of the differences in altitude, the colors often change at different times, causing a layered effect on mountainsides. Sometimes the top of mountains can be bare of leaves while the lower areas have just begun to turn. The winter months are mild, though recent years have witnessed more frequent storms. While temperatures can be low, snowfall is generally light. Winter has the advantage of clearing the trees of leaves and exposing ruins and structures normally obscured from view by vegetation. Elevations are higher in the southern Big Sandy River Valley and can cause cooler temperatures. In higher elevations, there can be ice patches and light snowfall on cool winter days.

Agriculture and Settlement

The Big Sandy River Valley has never had extensive agriculture. Crops were grown and consumed locally. The most common have been corn, wheat, hay, oats, potatoes, sorghum, and rye. Apples and peaches are produced in some areas. Travelers will see many small cultivated patches of land alongside homes and occasionally surplus crops for sale, but they will seldom find large farms in operation. Other crops that grow wild and are frequently harvested are blackberries, huckleberries, and ginseng, a root used for medicinal purposes.[13]

Unlike other areas of the greater Ohio Valley, the Big Sandy region has never been heavily populated. It is not clear when the first Native Americans inhabited the valley, but over the centuries many peoples have passed through the region. There were a few scattered Native American settlements in the north, near Louisa, Kentucky, but the valley was primarily used for hunting, especially by Shawnee and Cherokee. The arrival of Europeans and Americans in the eighteenth century brought a steady trickle of settlers to the area. By the early nineteenth century, scattered farming communities appeared on the flatlands alongside the rivers and creeks. Families survived by growing assorted crops, raising fowl and livestock, and hunting.[14] Cattle, swine, sheep, and horses were common, and hunting and fishing supplemented their diets. The nineteenth-century farm usually consisted of a log cabin with a chimney and porch made from locally available materials such as pine, stone, and assorted hardwoods. These were informal communities where people bartered and exchanged goods.[15] People occasionally visited local trade centers where they could exchange surplus crops and animal hides. Frame houses began to appear in the 1880s and were more common after the coal boom of the early twentieth century. As riverboats and railroads opened the region in the nineteenth century, opportunities increased.

The best land was usually located at the mouth of streams where they flowed into rivers. Along with small farms, commercial establishments such as mills and stores were built at these sites, as were schools and

East Kentucky Farmland, circa 1940. Works Progress Administration Collection, Public Records Division, Kentucky Department for Libraries and Archives.

churches.[16] Later settlers moved farther upstream, up hillsides, and into remote hollows where the soil was less fertile. The least desirable and least accessible lands were farthest inland at the source of the waters, known as the "head of the hollow," an expression still used today to designate isolated areas. Today, these same settlement patterns remain.

Like farming, early coal mining took place along the rivers and streams. As these easy-to-reach seams were exhausted, coal operators began to develop more remote sites for mining. Coal camps appeared where coal seams were exposed and railroads tracks were laid to the site. When there were great coal reserves, companies cleared large areas of timber and created communities in the wilderness. They developed towns and transported the equipment, materials, and people needed. As coal companies expanded into increasingly remote areas, they created towns almost overnight. Settlements spread up hillsides and through narrow hollows that previously had little use.

After spending some time in the region, visitors to the Big Sandy Valley will notice that many farming and mining communities have interesting, unique names. When studying Kentucky place-names, Robert Rennick found that landscape features often determined the names of early settlements. The Big Sandy region is filled with communities named branch, creek, point, or mountain. They are often accompanied by the names of early settlers or prominent residents such as John's Creek or Harold's Point. In east Kentucky, settlers viewed their landscape in terms of the countless tributaries that flowed from the hillsides into the forks of the Big Sandy River. Travelers unfamiliar with the local terrain will be confused by this maze of drainage. Local residents often give directions by referring to creeks and streams rather than roads. A traveler may be told to go up Greasy Creek or down Mate Creek. Many coal and railroad towns were named after company presidents, board members, supervisors, or engineers. For example, Jenkins, Kentucky, was named for Baltimore financier George C. Jenkins of the Consolidation Coal Company in the 1910s. Holden, West Virginia, was named for Albert Holden, a major investor in the U.S. Oil Company. Price, Kentucky, was named for Jack Price, a supervisor at Inland Steel's mining operations at Wheelwright, Kentucky.[17]

Transportation

The Big Sandy Valley's creeks and streams provided a good transportation system for early settlers. The banks served as footpaths for Native Americans and early pioneers, and flatboats, or push boats, were some of the earliest local river transportation. By the mid-nineteenth century steamboats opened the Big Sandy Valley to commercial traffic. Unlike flatboats that floated north to the Ohio River, steamboats could travel in both directions. They increased trade in small regional commercial centers and gave isolated family farms access to markets. The Levisa Fork had more areas that could be navigated than the Tug Fork, and by the 1870s there was regular steamboat service on the Levisa Fork between Pikeville and Catlettsburg, Kentucky, a distance of 120 miles.

Although rivers were used as a way to transport coal to northern markets, it was the arrival of the railroads that opened the valley. In fact, the great coal boom of the early twentieth century would not have been

The steamboat *Andy Hatcher* docked along the Levisa Fork of the Big Sandy River, circa 1894. Courtesy of Appalachian Photographic Archives, Alice Lloyd College.

possible without rail transportation. Three major railroads developed the southern Appalachian region: the Chesapeake & Ohio, the Norfolk & Western, and the Louisville & Nashville. The Interstate and the Clinch-field Railroads helped with the opening of southwest Virginia. Countless small independent spurs were constructed to connect remote mining sites with main railroad trunk lines. In many cases, coal companies laid tracks from their mines directly to the major rail lines. Early railroads in the Big Sandy Valley followed the narrow bottomland alongside creeks and rivers whenever possible. But the discovery of remote coal seams made railroad and coal companies more aggressive, blasting through mountains and building trestles to reach isolated areas. These railroads had a great impact on the design and layout of many coal towns. By 1920, the vast majority of communities in eastern Kentucky, western West Virginia, and southwest Virginia were linked to a major railroad.

The Chesapeake & Ohio Railroad was one of the most important. In the 1860s, Collis Potter Huntington held controlling interest in the Southern Pacific Railroad but dreamed of building a transcontinental line connecting the east coast of Virginia with San Francisco, California. He gathered a group of New York investors and purchased a small line, the Chesapeake & Ohio Railroad. By the 1870s, the line cut across West Virginia, reaching the new city of Huntington. The company was soon overextended, the western connections were never finished, and Huntington's dream of a transcontinental railroad failed to materialize. In 1876, the company was in trouble. It was reorganized a few years later and concentrated more on hauling lumber, agricultural goods, and coal to the east. In the 1880s it helped open the New River coalfields of West Virginia and built a new line to the port of Hampton Roads at Newport News. The Chesapeake & Ohio began to be an important supplier of coal to northern and eastern markets by the 1890s. By then it had competition from the Norfolk & Western Railway, which laid track into some of the same coalfields, but south of the C & O lines.

During the early years of the twentieth century the Chesapeake & Ohio opened the Big Sandy River Valley coalfields. Under the presidency of George Stevens, it acquired many smaller branch lines and helped develop mining in east Kentucky and West Virginia. In 1905, with the acquisition of the Ohio & Big Sandy Railroad, the Chesapeake & Ohio extended its tracks into the Elkhorn coalfields, terminating at Elkhorn City, Kentucky.[18] Within a few years it opened the Logan coalfields of West Virginia, made connections to the Hocking Valley Railroad, and acquired a direct line to Chicago. After building a large terminal at Russell, Kentucky, the C & O began to serve the Ohio and Great Lakes markets. Westward shipments overtook the eastbound flow.

The C & O survived the Great Depression because of the continuous need for coal. During the Second World War it became the largest coal hauler in the nation, with the world's largest railroad classification yard located at Russell. The company continued to thrive in the 1950s with further mergers and acquisitions. In 1972 it controlled the old Baltimore & Ohio and the Western Maryland railroads, and all of these railroads were merged into the Chessie System. Chessie later merged with Seaboard System and formed CSX Transportation. Today, most of the former Chesapeake & Ohio lines are part of CSX Transportation. These tracks wind through east Kentucky, often paralleled by U.S. 23 and U.S. 119.[19]

In 1881 several failing lines in the state of Virginia were consolidated to form the Norfolk & Western Railroad. The company's new owners built a line west into the Shenandoah Valley to a junction at Big Lick. Big Lick was later named Roanoke when it became the hub of the Norfolk & Western's entire operations. It was the site of the company's headquarters and its locomotive and freight car shops. The main line connected the port of Norfolk, Virginia, with the western gateway cities of Cincinnati and Columbus, Ohio. Branch lines reached into North Carolina, Maryland, and western Virginia. During the 1880s, as the railroad extended tracks into West Virginia and southern Virginia, it developed the mining town of Pocahontas and moved north through the Big Sandy River Valley, eventually reaching the Ohio River, connecting Bluefield with Kenova, West Virginia. The route followed the east bank of the Tug Fork, but spurs reached across the river into sections of Pike County, Kentucky. The one hundred mile stretch between Bluefield and Williamson, West Virginia, was the most challenging. The line crossed Elkhorn Mountain and then followed the twisting paths of Elkhorn Creek and the Tug Fork. Nineteen tunnels and a vast array of bridges were required to complete the line. In the first half of the twentieth century, the Norfolk & Western used electric locomotives along this section, but the completion of the 7,100 foot Elkhorn Tunnel in the 1950s allowed for the use of standard steam locomotives.[20]

The Pocahontas Division expanded and covered a broad area of West Virginia, southwest Virginia, and parts of east Kentucky. The main lines appeared like a capital letter "A" lying on its side. The route down the Tug Fork formed one side; a one hundred mile line from Bluefield west to Norton, Virginia, formed a second; and a forty-five mile crossbar ran from Cedar Bluff, Virginia, to Iaeger, West Virginia, to form a third. The route to Iaeger provided a shortcut for shipments north from the Virginia coalfields. These lines allowed for connections to the Interstate Railroad, the Clinchfield Railroad, and Louisville & Nashville. The last major extension of the line cut westward into Wise County, Virginia, reaching Norton in 1891 and opening the Big Stone Gap coalfields near Stonega. Countless spurs provided access to remote mine sites and company towns. The Pocahontas Division was physically challenging but ultimately very profitable. It accounted for more than 90 percent of the Norfolk & Western's coal traffic.[21]

The company experienced financial problems in the 1890s as it over-developed the coalfields and purchased two railroads in the west. In 1895 it declared bankruptcy but returned as the Norfolk & Western Railway. When it reemerged it became a premier railroad, unparalleled in profits and organization. It thrived in the early decades of the century, survived the Great Depression, and was the nation's most successful railroad by the 1950s, providing freight and extensive passenger service. Norfolk & Western common stock was an unrivaled blue-chip investment. The Pocahontas Division delivered countless thousands of tons of coal to coastal and Great Lakes outlets.

The Norfolk & Western gained a following among rail fans, as well as investors and coal operators. They designed their own locomotives, producing many unique electric and steam engines. By the 1930s, the Roanoke shops built special equipment to handle the heavy coal loads and steep grades of its tracks, especially in the Pocahontas Division. Well into the 1950s they continued to use massive, powerful steam locomotives that had long been abandoned by other lines. The Norfolk & Western was the last major railroad in the United States to change to diesel-powered engines. Some of its majestic engines and the scenic countryside they roared through were documented by people lamenting the passing of the great era of steam power, especially the O. Winston Link series on ghost trains.[22]

After the conversion to diesel, the railroad continued to thrive. A series of mergers in the 1960s expanded its freight service throughout the northeastern United States. In 1982 the Norfolk & Western merged with the Southern Railway and became part of the Norfolk Southern Corporation. In 1990 the Norfolk & Western lines were renamed the Norfolk-Southern Railway with a familiar "NS" logo. The Norfolk-Southern lines still serve much of West Virginia and southwestern Virginia. Today, the main tracks of the old Norfolk & Western continue to wind through the Big Sandy River Valley paralleling U.S. 52 and serving as the major transporter of coal through the region. [23]

The third major railroad to penetrate southern Appalachia was the Louisville & Nashville, which passed through Pineville and Middlesboro, Kentucky, in the 1880s. From the Cumberland Gap, the line extended into Lee and Wise Counties, Virginia, reaching Norton in 1891 and connecting with the Norfolk & Western. Numerous rail spurs led to mining areas along these routes and gave coal companies access to markets in both the Ohio and Mississippi valleys. These three major railroads shaped the development of the Big Sandy region in the twentieth century and helped create many of the coal towns described in this guidebook.[24]

Two other railroads opened the coalfields of southwestern Virginia. In 1882 the Virginia Coal and Iron Company was organized, and it purchased thousands of acres of land in southwest Virginia. By 1895 it opened mines in Wise County and built a rail line to haul coal from the mines. The next year this line was organized as the Interstate Railroad. It shuttled loaded coal cars between the mines and Appalachia, where it met the South Atlantic & Ohio Railroad. Because of a dispute with the Louisville & Nashville Railroad in 1903, the Interstate was greatly expanded. Over the next twenty years it reached into the many coal camps developed by Stonega Coke and Coal Company, a lessee of Virginia Coal and Iron.

The Interstate absorbed several small lines and dominated rail traffic in the mines around Wise County during much of the twentieth century. A railroad yard and service facilities were built at Andover. Other yards connected with the Louisville & Nashville, Norfolk & Western, and Southern Railroads. In 1961 Interstate became a subsidiary of the Southern, and in 1985 it was reorganized as part of the Norfolk-Southern Railroad.[25]

The final phase in opening the coalfields of the Big Sandy Valley was the completion of the Clinchfield Railroad. In 1902 George L. Carter and New York banker James A. Blair organized the South & Western Railway in Virginia. Two years later they established the Clinchfield Coal Corporation and acquired more than four hundred thousand acres of land in the area. The South & Western Railway was reorganized as the Carolina, Clinchfield, & Ohio Railway and began building a line from Spartanburg, South Carolina, to Elkhorn City, Kentucky, where it eventually connected with the Chesapeake & Ohio. In 1915 this final rail line opened up the vast coal reserves at the headwaters of the Big Sandy River and allowed for the transport of coal south to the textile mills in the Carolinas and north to the Great Lakes.[26]

Markets

Once the coal was funneled into railroad hoppers at the tipple, it was transported to regional marshaling yards, where the cars were inventoried, sorted, and forwarded to coal dealers and brokers. Coal cars were then taken to steel mills or power plants, where they were dumped, forming massive mountains of coal. Large factories and plants had their own tracks, trestles, and locomotives to sort and unload coal shipments. The coal was sometimes dumped out of the hoppers into piles and trucked to the final destination. Much of the coal was taken directly to ports for shipment north. The Norfolk & Western piers at Norfolk, Virginia, and the Chesapeake & Ohio complex at Newport News received much of the

The first locomotive arrives in Jenkins, Kentucky, 1912. Jenkins Collection, Mary Jo Wolfe Memorial Library.

coal from the Big Sandy River Valley throughout the twentieth century. The coal was dumped directly into ships that carried it north to east coast and Great Lakes cities.[27]

Roads

The Interstate system bypasses most of the Big Sandy River Valley, but a series of federal, state, and county highways cross through the region, and all of the sites included in this guidebook can be reached from these routes. Three federal highways pass through the greater Big Sandy River Valley: U.S. 23, U.S. 52, and U.S. 119. Since travelers may enter the region from any direction, a number of sites related to the coal industry that lie beyond the Big Sandy Valley have been included and are easily accessible from these federal highways. Many state and county highways penetrate the area and link towns that were previously isolated. These roads connected with the major federal highways and have created a fairly complex road network by the late twentieth century. In this guidebook, local roads are labeled with state or county route numbers such as KY 102 or VA 78 even though many have proper names such as Derby Road or Haymond Road. The route numbers will always be indicated and secondary names will be mentioned where they are useful. In order to understand the history of the Big Sandy area after the coal boom of the 1910s, it is important to briefly trace the development of three major transportation arteries: U.S. 23, U.S. 119, and U.S. 52. The First Federal Highway Act of 1921 designated a number of local and regional roads to be joined together as federal highways. The construction of these highways occurred at a time when many Big Sandy coal towns were at their peak. These roads led into isolated settlements that before could only be reached by rail. In fact, the highways often paralleled the main railroad lines. Unlike today's interstate system, they connected small towns and cities, often passing directly through downtown commercial districts. Today they offer travelers a great way to experience the Big Sandy region. These roads wind through countless hollows, crossing creeks, rivers, and railroads. Along the roadside travelers can see many reminders of the region's coal mining past: abandoned tipples, railroad grades, and company houses, offices, and stores. Many of the roadside views are spectacular as the highways descend into river valleys and climb through the mountains.

U.S. 23

The Kentucky section of U.S. 23 was originally known as the Mayo Trail, named after John C.C. Mayo. In the 1910s it was an unpaved route linking Paintsville with Jenkins. During the 1920s, U.S. 23 was graded and new sections were opened. By 1932 it connected lower Ohio with Tennessee and became known as the Appalachian Way. Under the Works Progress Administration (WPA) of the 1930s, U.S. 23 continued to be improved and paved, reaching Florida by 1940.[28] When complete, it connected Detroit, Michigan, with Jacksonville, Florida, and became one of the main north-south highways in the eastern United States. Although Interstate 75 in central Kentucky has replaced U.S. 23 as the main route from the Great Lakes to the South, U.S. 23 is still an important highway in east Kentucky and southwest Virginia. It is continually being improved, rerouted, and widened.

When U.S. 23 was first finished, planners hoped it would open up the Appalachian region to greater development and economic diversification.

U.S. 23, a modern four-lane highway, passes through Pikeville, Kentucky.

Ironically, by the 1950s it became a pathway out of the region for unemployed miners and their families heading north for industrial work. Over one million people left Appalachia after 1950, and U.S. 23 became a convenient means of escape, a road to the factory jobs of Ohio and Michigan. In the words of country music artist Dwight Yoakam, the way to succeed in eastern Kentucky was to learn "readin', writin', and Route 23."

Today, much of U.S. 23 is a four-lane highway with many divided sections. Flat, open areas with minimum grades were first widened and resurfaced in the 1970s. Certain four-lane stretches have been rerouted, and sections of old U.S. 23 parallel the highway. Some of the coal mining and railroad sites described here are no longer located along U.S. 23, but the road provides a good basic route along the Levisa Fork and the CSX tracks. U.S. 23 has also been promoted as Kentucky's "Country Music Highway" because of the many recording artists and entertainers that came from communities along its path through the state. Dwight Yoakam, Loretta Lynn, Ricky Scaggs, Tom T. Hall, Patty Loveless, and Crystal Gayle all trace their roots to the Big Sandy region. Unfortunately, driving along this road at sixty-five miles an hour might cause travelers to lose sight of how remote and isolated this part of the country was until just a few years ago. In some places, old U.S. 23 has been reclassified as a state or county route. Following the old road provides an opportunity to imagine life when these coal towns were booming.

U.S. 119

Federal highway 119 was begun in 1924 to link DuBois, Pennsylvania, with Pineville, Kentucky. The section connecting Williamson, West Virginia, with Pikeville, Kentucky, was built in the early 1920s and was considered an engineering marvel at the time of its completion. Pike County had long been divided by a mountain ridge separating the east and western portions of the county. Travel between the two areas was limited, keeping eastern Pike County quite isolated. Western Pike County, along the Levisa Fork, developed as the Chesapeake & Ohio Railroad reached south from White-house in the early twentieth century. Eastern Pike County developed as the Norfolk & Western Railway opened the Tug Valley of West Virginia. U.S. 119 ended that division, and a gala celebration accompanied its opening in 1924, with the governors of both West Virginia and Kentucky meeting at Bent Branch on John's Creek. A wedding ceremony was conducted with a West Virginia Hatfield marrying a Kentucky McCoy.[29] Today U.S. 119 has been greatly improved and almost all of its winding two-lane segments have been replaced with four lanes.

U.S. 52

U.S. 52 originates at the Ohio border and runs along the southwest boundary of West Virginia, often paralleling the Tug Fork. At Williamson, it crosses U.S. 119 and continues through southern West Virginia. U.S. 52 was constructed during the 1920s to connect Huntington and Williamson, West Virginia, and it generally paralleled the Norfolk & Western Railway. Prior to the construction of U.S. 52, most of the towns along the Tug Fork were accessible only by rail and riverboat. The northern Tug Fork was once the site of many small coal towns in the 1920s, but by the 1930s few remained active. Many of the homes, company buildings, and mining sites along U.S. 52 have been lost during the late twentieth century. In fact, it is possible to drive along vast stretches of the highway in Wayne and Mingo Counties without seeing any evidence of the region's coal mining past. Flood control projects have destroyed much of the historic landscape along the highway as well. Some communities, such as Borderland, West

Virginia, will no longer exist when these projects are complete. South of Williamson, U.S. 52 heads east away from the Tug Fork and the rails but still provides relatively easy access to the towns along the river through a series of connecting state and county highways. It crosses the mountains through Varney, Mountainview, and Hampden, meets the Guyandotte River at Gilbert, and heads back toward the Tug Fork at Iaeger. From Iaeger, U.S. 52 follows the river and railroad once again to Bluefield. U.S. 52 has not undergone the extensive widening and improvement that U.S. 23 and U.S. 119 have. It remains a challenging road at times but provides visitors with a route through the Big Sandy River valley that has not changed much over the years. The landscape along this older highway remains full of railroad tracks, mining sites, small commercial centers of towns, and houses. Like some of the older stretches of U.S. 23 and U.S. 119, U.S. 52 is a great way to tour the region at a more leisurely pace or experience these towns more as they were fifty years ago. The southern part of U.S. 52 in West Virginia has been designated as part of the Coal Heritage Trail, a trail connecting many historic mining communities. Heritage tourism projects are developing this segment of U.S. 52 as an important tourist destination. There are currently plans to build a ninety-mile King Coal Highway connecting the Williamson area with Bluefield and the Interstate 77 interchange. Depending on the route of this new road, large sections of U.S. 52 and the communities it serves could be adversely affected.

These three major federal highways—U.S. 23, U.S. 119, and U.S. 52—guide visitors through the region, taking them through the heart of the river valley and to countless coal towns and mine sites along the way.

Map

The map in this guidebook offers the traveler an overall view of the Big Sandy River Valley. It is not to scale and includes only the main roads and select coal towns. In order to keep the map legible and easy to read, many features have been left out. Before traveling through the Big Sandy region, it is a good idea to obtain a state roadmap from tourist information centers or state highway departments that provides updated and more detailed information. Almost all of the towns and roads discussed are easily accessible from federal and state highways, but for exploring remote sites more precise maps may be needed. County courthouses can usually provide county highway maps with more detailed information. Excellent maps can also be found in the DeLorme series of state atlases. There are individual atlases and gazetteers for the states of Kentucky, Virginia, and West Virginia that show roads, railroad lines, rivers, creeks, and terrain. There is a grid system for locating sites, mileage systems, and information on Global Positioning Systems (GPS). Most small settlements and landmarks mentioned in this guide are indicated on these maps.[30]

For more intense study, topographic maps prepared by the U.S. Geological Survey may be useful. Each map covers a seven-by-nine-mile area and shows roads, buildings, mine sites, and landmarks. Each also indicates the contour or slope of hillsides and elevations. College and university libraries are the best places to consult these maps, and many have complete collections that cover the entire state and bordering regions. The University of Kentucky at Lexington, Marshall University in Huntington, and the University of Virginia's College at Wise all have extensive collections. Limited collections can also be found at Alice Lloyd College in Pippa Passes, Kentucky, Morehead University, and some community college libraries. It is best to make arrangements ahead of time to use these materials. Current individual maps can be purchased directly from the U.S. Geological Survey in Denver, Colorado; state geological survey offices; or bookstores and map vendors in the Big Sandy region.

These maps, frequently updated to provide an accurate picture of current conditions, can provide a wealth of information about a site. Older maps, which can be viewed in libraries and archives collections, provide valuable historic information about conditions before coal companies ended operations. Former building locations, roads, mine sites, and railroad lines can be located. By studying these maps, one can appreciate the difficulty of settling such steep terrain and narrow valleys.

Coal Mining Communities

An extremely important part of the Big Sandy River region is its people. They came from throughout greater Appalachia and many other parts of the world. Once settled in these coal towns, they developed a closeness and sense of place that was reinforced by an isolated location and dependence on coal companies for everyday needs. Many stayed for generations, long after the coal companies abandoned these towns. When people did move they tended to stay close by. Many went to nearby coal towns where work was available. It was only in the 1950s, with the collapse of the coal industry, that out-migration took a serious toll. Even so, many former residents continued to maintain close ties to these communities. These coal towns have held people together and created a sense of community, leading to many efforts to preserve the coal mining heritage of the Big Sandy River Valley. A sense of community is evident in many activities that have developed in the late twentieth century.

One example has been the growth of reunions and homecoming celebrations in coal towns in recent years. Several places in the Big Sandy region now have regularly scheduled celebrations that draw people from faraway towns and cities. Many of these center around a holiday or anniversary and allow people to enjoy entertainment, visiting, local foods, as well as reminiscing about a coal town's heyday. Seco, Kentucky, has a Miners' Reunion in the beginning of May, and Appalachia, Virginia, celebrates Coal and Railroad Heritage Days in August. Pikeville's Hillbilly Days in mid-April is one of the largest. Many coal towns have family and community reunions on a regular basis. Some of these communities have developed historical societies, commissions, and offices to promote and preserve their coal mining heritage. The Jenkins (Kentucky) Heritage Commission and the Van Lear (Kentucky) Historical Society have been particularly active in preserving mining heritage in their communities. In recent years, some of the existing agencies and historical societies, such as the Wise County Historical Society in Virginia and the Tug Valley Chamber of Commerce in West Virginia, have taken a greater interest in the coal mining past.

Another example of community effort has been the rather recent publication of books, pamphlets, and calendars that portray coal mining communities during their prime years. Many of the people who made contributions to this guidebook are part of a small but dedicated group who have kept the memories of these towns alive, despite their communities' rapid decline in the late twentieth century. Community histories have been published that usually include an overview of the area, the development of individual coal mining towns, detailed family histories, and many illustrations. Many of these have been the first publications about these towns and have provided a useful source for basic information about their development.

A by-product of these efforts has been the preservation and restoration of many buildings that were scheduled for demolition or had been abandoned over the years. In some coal towns, former depots, company stores, or residences now serve as museums or centers of information. In other communities entire sections are being restored and developed into tourist sites. In the Big Sandy region the best examples are Lynch and Benham, Kentucky. Other locations include Big Stone Gap, Virginia; Coalwood and Matewan, West Virginia; and Seco, Jenkins, and Stone, Kentucky.

A great deal of cooperation has developed between local organizations and state and federal agencies. In fact, entire districts of the Big Sandy region have been surveyed to determine their potential for preservation and heritage tourism projects. In the 1990s, the Kentucky Heritage Council and the West Virginia Division of Culture and History helped identify sites, targeting structures for preservation and surveying areas to determine their potential for heritage tourism projects. An eleven-county area of West Virginia that includes Mingo, McDowell, Logan, and Mercer Counties was designated as a National Coal Heritage Area, and projects are now under way to prepare coal mining areas for tourism and to produce interpretive materials to guide visitors through the area. In this same region, plans are in progress for a King Coal Highway that will make travel easier, safer, and more convenient for tourists. In addition, Matewan's downtown commercial district has been designated as a National Historic Landmark. The Kentucky Heritage Council has also been active in adding coal mining sites to the National Register of Historic Places and helping develop local tourism projects. Recently, the C. B. Caudill Store in Letcher County was added to the National Register of Historic Places. Sites at Benham and Lynch continue to be developed as well, and more historic markers are appearing in these towns. No doubt this work will make it easier for travelers to appreciate the area's coal mining heritage.[31]

The coal towns of the Big Sandy region, despite their decline, continue to have active communities that contribute greatly to historical preservation and heritage tourism sites. Visitors will find a great variety of structures remaining throughout the region and will encounter local groups of volunteers working to preserve the coal mining past of their communities. Although the coal towns discussed in this work vary in structural integrity and degree of preservation, all offer insight into the coal mining past of the Big Sandy River Valley.

A General History of the Big Sandy River Valley

2

Native Americans in the Valley

The Big Sandy River Valley has been inhabited since prehistoric times and has played an important role in the shaping of modern American history. Around 5000 B.C., Native American hunters began camping along the rivers and streams of the region. By 1000 B.C. Adena Mound Builders lived in the Ohio River Valley, surviving by mixing agriculture, food-gathering, hunting, and trade. They also entered northern parts of the Big Sandy Valley. The Mound Builders disappeared by about 500 A.D., but by the historic period, several Woodland tribes had settled the region. The Totero, or Chatteroy, Indians are believed to have occupied the Big Sandy Valley in the seventeenth century, attracted by bountiful game, rich soil, and timber. By the early eighteenth century, Shawnee, Mingo, and Cherokee hunters roamed the area during their winter hunts, but none permanently settled here. By the eighteenth century Native American trappers were involved in the European fur trade and may have pushed smaller tribes like the Totero out of the region. Despite thousands of years of Native American activity, visitors will see little of these cultures remaining today. Four Adena mounds once stood near Paintsville, Kentucky, but were destroyed by railroad expansion in the 1940s. Some newly discovered mounds, such as those at Cotiga, West Virginia, are currently being excavated.[1] Only a few place-names, such as Chatteroy, West Virginia, or Mingo County, reflect the Native American heritage of the valley. U.S. 52, which parallels the Tug Fork of the Big Sandy, was once an important trail for Native Americans connecting southern Ohio with the Valley of Virginia.[2]

Arrival of the Europeans

During the seventeenth century, Europeans began to reshape life along the Big Sandy. The French moved south from Canada to trade with the Shawnee, and English merchants from New York and South Carolina developed a lively exchange with the Iroquois and Cherokee. No doubt some of these traders entered present-day West Virginia and east Kentucky,[3] and it was probably a seventeenth century fur trader who first traveled through Big Sandy River Valley. Surveyors from land companies soon followed. In 1750 Dr. Thomas Walker of the Loyal Land Company of Virginia led a party of explorers through southwest Virginia and the Cumberland Gap into east Kentucky, moving in a northeasterly direction. By June 1750 they were in the Big Sandy Valley. The party moved down Big Paint Creek until they arrived at the present site of Paintsville, where Walker named the river at the mouth of Paint Creek the Louisa River. It later became known as the Levisa. Less than two weeks later, Walker was on the Tug Fork and named it Laurel Creek. After his four-month trek through the rugged terrain, he had accumulated a great wealth of knowledge about the Big Sandy River Valley, including the locations of several rich seams of coal.[4]

Walker's Loyal Land Company was in competition with a second Virginia real estate firm, the Ohio Company. Shortly after Walker returned, Christopher Gist led a party of explorers down the Ohio River to pave the way for a colony. They floated as far west as the mouth of the Great Miami River before abandoning their canoes and traveling overland across Kentucky. Gist eventually crossed Pound Gap (along the path of U.S. 23 in Letcher County) in spring 1751.[5] Although the French and Indian War (1754–60) disrupted the plans of both companies, the reports of Walker and Gist brought attention to the valley. By the 1760s, well-known Long Hunters like Daniel Boone, Michael Stoner, and members of the Skaggs family were in the Big Sandy Valley. Daniel and Squire Boone left their homes in the Yadkin Valley of North Carolina to hunt in the Kentucky country. They traveled along the Russell Fork of the Big Sandy north, through the "Breaks" and into eastern Kentucky. They followed a buffalo trace into Floyd County, camping at present-day David. At first, the Long Hunters tried to blend into the valley, living a life similar to that of the Indians, but after a while conflicts developed. As more Europeans arrived, tensions increased and Indians were forced to leave the valley. Even so, the Shawnee continued the tradition of the winter hunt into Kentucky until the 1790s.

Indian Conflicts and Jenny Wiley

By the time of the American Revolution, the bluegrass area of central Kentucky was being settled. White farmers, many with slaves, gathered around present-day Lexington. After the war, another wave of migrants crossed the Appalachian Mountains and built homesteads in the eastern and western portions of the Kentucky District of Virginia, but few settled in the Big Sandy region. One pioneer who settled in the John's Creek area of Kentucky was Tice Harman. He regularly hunted in the area and is generally credited with founding the first permanent American settlement in the Big Sandy River Valley. The Harmans were in almost constant conflict with various Indian groups. After one bloody encounter during their fall hunt in 1789, a group of Indians retaliated by attacking the cabin of Thomas Wiley, located along Walker's Creek in Bland County, Virginia. Apparently, Thomas Wiley was not at home when the war party attacked on 1 October 1789, killing his brother-in-law and three of his children. His wife, Jenny, and their infant child were taken by the Indians into the Big Sandy Valley. Two rescue parties were quickly organized: Thomas Wiley joined a party that pushed into the New River Valley while Harman led a second group into the Big Sandy Valley. Near the mouth of John's Creek, near present-day Auxier, they erected a block-house fort. Because of high waters, the Indians could not return to the Shawnee villages north of the Ohio River and instead camped along Little Mud Lick Creek. In spring 1790, Jenny Wiley escaped from her captors and made her way to Harman's Station. She returned to live with her husband at Walker's Creek until 1800, when the Wiley family moved to a new homestead at River, Kentucky.[6] The Jenny Wiley story is well known throughout the valley, and several sites bear her name. Jenny Wiley State Park is located off U.S. 23 near Prestonsburg, Kentucky. Travelers can visit Jenny Wiley's grave near the town of River, in Johnson County Kentucky, off KY 3224 north of Prestonsburg.

Americans Come to the Valley

About the same time that Harman's Station was being built, other scattered settlements developed. Charles Vancouver organized a settlement at present-day Louisa. To the south, the Leslie family moved onto land along Brushy Fork in Pike County, and John Spurlock built the first cabin at Preston's Station, now known as Prestonsburg. Because of the irregular settlement of the region, conflicting land claims soon became a problem. Early surveyors marked off large tracts of land using vague, natural boundaries. Overlapping claims caused endless confusion.[7] Even today insecure land titles continue to provide lawyers of the Big Sandy Valley a lively business.[8]

Most Americans moving into Virginia's Kentucky District continued to pass by the Big Sandy Valley heading for the bluegrass region along two major routes.[9] Daniel Boone's Wilderness Road brought settlers up through modern southwest Virginia to the Cumberland Gap (today VA 58 parallels much of this route). The Ohio River was the main route for settlers coming from western Pennsylvania and Maryland.[10] By the 1790s a few hardy individuals and families made their way to the Big Sandy River Valley, and by the early nineteenth century independent farmers owned much of present-day east Kentucky, southern West Virginia, and southwest Virginia.[11] This rugged group of pioneer settlers became known as mountaineers.

By the 1810s, although lightly populated, a rich Big Sandy River Valley culture had emerged. The pioneer settlers of the valley were typical of those throughout the western American frontier. English and German families mixed with Scots-Irish and a few French Huguenots. Today, many of the region's common family names reflect these origins. Local tax rolls are dominated by Anglo-Saxon names such as Hatfield, Williamson, Hunt, and Johnson. A considerable number of Scots-Irish names, such as McGinnis, McCoy, McLanahan, McKenzie, Wiley, and Swiney, are also present. French and Anglo-Norman names are also common, but their spellings have often been changed. Mollett, Gullett, Stumbo, Stambaugh, Marsillett, and Maynard are all of French origin. Names such as Harman, Marcum, and Crider are probably Anglicized versions of German names.[12]

There is also a degree of religious diversity in the valley. Organized religions were slow to develop, so revivals sparked many new conversions in the early nineteenth century. As in much of the southern backcountry,

Baptist and Methodist denominations became the most common. Prominent denominations in the mountains include Old Regular Baptists, Primitive Baptists, Free Will Baptists, and Pentecostals. These religions added emotion as an important part of the worship experience. While their theologies differ, they all emphasize individual conversion, adult baptism, and the authority of the Bible. Until the 1940s, many mountain congregations were served by untrained preachers who received no regular pay for their duties. Today, travelers will see many of these churches throughout the region.

Although the Big Sandy Valley was isolated, basic transportation systems did develop. Good trails and roads were rare, making rivers and creeks the best means of travel. Canoes and push boats moved up and down both forks of the Big Sandy from the earliest days of European settlement. By the 1830s, steamboats traveled the main channel of the river as far as Louisa, and in 1842 Allen Hatton guided the first steamboat south along the Levisa Fork.[13]

An early-twentieth-century log home in the Big Sandy River Valley. Works Progress Administration Collection, Public Records Division, Kentucky Department for Libraries and Archives.

The Civil War briefly interrupted the development of the Big Sandy region. Virginia joined the Confederacy, but its western counties remained loyal to the Union and became the new state of West Virginia. Kentucky was a slave state but remained in the Union. Even so, some areas actively assisted the Confederacy and organized local militias to combat Union troops. Other areas supported the Union cause. Though the percentage of slaveholding families was small in east Kentucky, there was a significant slave population. Most residents of the valley did not own slaves, but they did support the slave system.[14] As a result, there was considerable guerilla activity and several skirmishes took place between Union and Confederate forces. An 1861 encounter took place at Ivy Mountain on the road between Prestonsburg and Pikeville, Kentucky, where Union Gen. William O. Nelson defeated Confederate troops. In 1862 Confederate Brig. Gen. Humphrey Marshall reentered the valley through Pound Gap and led his troops north to Hager Hill in Johnson County. Union Col. James A. Garfield's forces met them at Middle Creek near Prestonsburg. Garfield continued south and regained control of the valley, establishing his headquarters at Pikeville. Marshall returned in 1863, and John Hunt Morgan also passed through the region during his last raid into Kentucky in June 1864. In the Big Sandy Valley there were many irregular forces and guerrilla raids, often controlled by mountain families. According to historian Carol Crowe-Carraco, "conditions in the mountains defied description. Murder, robbery, pillage, plunder, and starvation plagued the Big Sandy."[15] Future feud leaders William Anderson "Devil Anse" Hatfield and Randolph "Ranel" McCoy rode with a Confederate partisan band known as the Logan Wildcats.[16]

Travelers in the Big Sandy Valley can visit several Civil War sites. The Jenny's Creek battlefield is located where the Turner house now stands at the junction of the new section of U.S. 23 and U.S. 460, just west of Paintsville, Kentucky. The Middle Creek battlefield is located about three miles west of Prestonsburg at the junction of KY 114 and KY 404. A historical marker indicates the site of the battle, which is occasionally reenacted by local Civil War enthusiasts. The city park in Pikeville was the site of Garfield's headquarters, and several points along U.S. 23 in western Pike County are marked as part of the Pound Gap campaign. The Battle of Ivy Mountain was fought near the point occupied today by the Wagon Wheel Restaurant in Ivel, Kentucky. Most of the battlefield was destroyed when U.S. 23 was widened.

Steamboats, Timber, and Feuds

Shortly after the war, steamboat traffic on the Levisa Fork expanded at a phenomenal pace. Packet lines were organized during the 1870s carrying passengers and freight from Pikeville to Catlettsburg. Efforts were also made to improve navigation on both major forks of the river by pressuring the federal government to appropriate money to remove obstacles and build needed dams.[17] Near the mouth of the Ohio River a thriving iron industry had developed by midcentury. Iron furnaces were located throughout the Hanging Rock region and some limited coal mining was taking place in Carter and Boyd Counties in Kentucky.

Logging became a big business in the Big Sandy region. During the fall and winter months, trees were cut and formed into rafts that were floated down the river to Catlettsburg, where they were sold to timber brokers.[18] During the 1890s and into the early years of the twentieth century, countless thousands of logs were floated down the forks of the Big Sandy each year. Catlettsburg often resembled the cattle towns of the Old

The Hatfield Clan in 1897. "Devil Anse" Hatfield is seated in the center. Norfolk & Western Historical Photograph Collection, Virginia Polytechnic Institute and State Universities Libraries.

West as each spring young loggers drew their pay and spent their money on whiskey and women. Fights broke out and occasional shootings took place.[19] Unfortunately, trees were cut with little concern for future growth, and forests were quickly depleted. The timber trade also revived old arguments over land claims and led to a series of feuds. The most celebrated of these conflicts, well known throughout the valley, was the Hatfield-McCoy Feud fought in Pike County, Kentucky, and Logan County, West Virginia.

Railroads and Coal

As the logging business boomed, a new coal industry emerged. The first large-scale coal mining operation began at Peach Orchard in Lawrence County in 1847 when the Peach Orchard Coal Company of Cincinnati purchased land south of Louisa. Within a short period of time a camp was constructed. This was the first company coal town in the Big Sandy Valley.[20] Coal mining at Peach Orchard was interrupted by the Civil War but was revived in 1881 when George S. Richardson brought a short-line railroad to the site. It allowed the Great Western Mining and Manufacturing Company to open new mines on Nat's Creek, which led to the construction of New Peach Orchard. According to historian Henry P. Scalf, "it was New Peach Orchard that stood out as a model for the construction of a mining town."[21] In 1887 the Chatteroy Railroad was extended southward into Johnson County, all the way to Whitehouse.

The coal boom at Peach Orchard was part of the overall development of southern Appalachia. Railroads were built to open up the vast coalfields to mining.[22] By the 1880s the Norfolk & Western Railway was laying track down the Tug Fork from Virginia to the Ohio River. In the area of Cumberland Gap, Alexander Arthur and his American Association, Ltd., attempted to establish a coal and iron empire. Arthur's Middlesborough scheme collapsed in 1891 and the depression of 1893 slowed development, but as the Big Sandy region entered the twentieth century it was poised for dramatic change. Coal production would quickly replace farming and timber cutting and would soon dominate the local economy.

Southwest Virginia experienced similar changes, especially in Wise County. By the late 1870s Rufus Ayers, a prominent Scott County lawyer, had accumulated more than ten thousand acres of coal lands near present-day Big Stone Gap and had invested in railroads, iron, and coal. Ayers managed the Virginia Coal and Iron Company and helped establish the first coal mining operations at Stonega. George L. Carter was even more successful than Ayers and expanded his industrial investments into Kentucky and Tennessee. In the 1890s he created the Carter Coal and Iron Company, which developed areas in Roanoke, Bristol, Big Stone Gap, and Middlesboro.[23] In east Kentucky, the most successful entrepreneur was Paintsville's John C.C. Mayo, who played a vital role in luring large mining conglomerates into the Big Sandy Valley. Using a legal device known as the "broad form deed," Mayo was able to make a fortune by persuading farmers to sell the mineral rights beneath their land while allowing them to retain the surface rights to their property. The mineral rights were later resold to large, out-of-state coal companies.

The Mayo mineral rights became particularly valuable after the Chesapeake & Ohio Railroad purchased the track line of the old Chatteroy Railroad and extended its line up the Levisa Fork of the Big Sandy River.[24] These broad-form deeds became common in the late nineteenth century, and, as Harry Caudill noted, they "authorized the grantees to excavate for minerals, to build roads and structures on the land and to use the surface for any purpose 'convenient or necessary' to the company or its successors

John C.C. Mayo Exploration and Inspection Party along Shelby Creek in Pike County, Kentucky. Courtesy of Appalachian Photographic Archives, Alice Lloyd College.

in title . . . The landowner's estate was made perpetually 'servient' to the superior or 'dominant' rights of the owner of the minerals." And for good measure, a final clause absolved the mining company from all liability to the landowner for such damages as might be caused "directly or indirectly" by mining operations on his land.[25]

West Virginia's Mercer and McDowell Counties also grew. They were situated on the Pocahontas coal seam, one of the richest in the country. When the Norfolk & Western Railroad reached the new coal town of Pocahontas, Virginia, in 1883, a great mining boom began. As the railroad slowly made its way northwest into West Virginia, small mining communities began to appear. A number of investors made quick fortunes in the Flat Top coalfield, and an entire community, Bramwell, became the home to these new millionaires. As the railroad reached north, it entered the headwaters of the Tug Fork of the Big Sandy River. By the 1910s large coal mining operations were in place at Gary, Elkhorn, Coalwood, and Caretta, and Bluefield became an important commercial and transportation center. As the railroad made its way north along the Tug Fork, Mingo and Logan Counties in West Virginia opened to development. A number of small railroad lines connected with the Norfolk & Western and with the Chesapeake & Ohio lines that came from the north and east. By the 1910s the east bank of the Tug Fork saw a rapid expansion of its coal mining industry as branch lines reached across the river into east Kentucky.[26]

The early years of the twentieth century also saw the rise of several large corporations that dominated the coal industry and played an important role in the development of the Big Sandy region. Small coal operators did not have the capital or technology to compete, so instead they sold or leased lands to these new industrial giants. Consolidation Coal Company (also known as Consol), begun in Maryland in 1864, entered the coalfields of Pennsylvania, Kentucky, and West Virginia. Among its early directors were Warren Delano, William Aspinwall, James Roosevelt, John Murray Forbes, and Erastus Corning, family names that would dominate American industry throughout the twentieth century. Consolidation Coal developed massive mining operations and employed thousands of people. By 1927 it was the biggest independent coal producing corporation in the United States. Subsidiaries and related companies also mined the Big Sandy Valley. A group of prominent West Virginia coal operators known as the Fairmont Ring included Clarence Wayland Watson, Aretus Brooks Fleming, James Edwin Watson, and Alpheus Haymond. They guided Consolidation Coal from 1903 to 1927 and organized the Elkhorn Fuel Company, a major mining company affiliate in east Kentucky. Virginia Iron, Coal and Coke, with the backing of the Roosevelts and Delanos, mined vast areas of the southern Appalachian coalfields. The Virginia Iron and Coal Company developed most of the coal towns in Wise County, Virginia, through its lessee, Stonega Coke and Coal. Island Creek Coal bought and leased coal lands throughout West Virginia, developing much of Logan County. Industries that were dependent on coal also leased lands and developed their own "captive" mines where they mined coal for their own uses. U.S. Steel created the company towns of Lynch, Kentucky, and Gary, West Virginia. Inland Steel bought Wheelwright, Kentucky, in the 1930s and used its mines as a source of coking coal. There were countless small operators in the region, but the big coal companies dominated the industry, wielding political as well as economic power. They influenced state legislatures, placed powerful lobbyists in Washington, and helped elect officials who would support and promote the coal industry. For example, Clarence Wayland Watson served as a U.S. senator from West Virginia and was chair of the board of Consolidation Coal. Leaders of large coal companies had the capital and the manpower to control coal producing states and influence the federal government, and they were often instrumental in protecting the coal industry from extensive regulation and unionization.[27]

By the second decade of the twentieth century, the transformation of the Big Sandy Valley was complete. Coal revolutionized both the economy and culture of the region as company coal towns stood in sharp contrast to the quiet farming settlements of an earlier day. Mountaineers left their farms and became coal miners, but the native workforce was not enough. Coal companies soon recruited workers from nearby states, and white and African American southerners came to work in the mines.

The coal boom became one of the few ways for blacks to enter the industrial working class, and for the first time many of these remote mountain communities had sizable African American populations. Blacks and immigrants, however, were usually separated from the native white population. In Kentucky and Virginia, African Americans were subjected to legalized segregation in all aspects of their lives. Although West Virginia did not have Jim Crow laws, segregation and racism were everyday facts of life. African Americans and whites had separate social, religious, and educational lives. Throughout the region, blacks attended separate schools, lived in clearly defined neighborhoods, and were relegated to separate

An ethnically diverse coal mining workforce at a meeting in Benham, Kentucky, circa 1920. Reprinted from *Coal Age* (1921).

areas of theaters, restaurants, and stores. Black coal miners were given dirtier, more dangerous jobs than whites and seldom held supervisory positions. Economically, coal mining in West Virginia gave many African Americans a chance to progress, even creating a substantial black middle class within their own communities, but socially they faced the same discrimination found in other southern states.[28]

Europeans also made up much of the workforce in the Big Sandy region, especially Welsh, Italian, Hungarian, and Slavic immigrants. While some coal companies hired only white, American-born laborers, others, such as Consolidation Coal, Elkhorn Coal, and Inland Steel, employed mountaineers, African Americans, and European immigrants. Before the First World War, thousands of immigrants made their way to the coalfields of the Big Sandy region. Some communities, such as Himlerville in Martin County, Kentucky, had large immigrant populations. Many other towns throughout the region had sizable numbers of immigrants from almost all parts of the world. However, immigration tapered off in the 1910s, and by mid-twentieth century much of the ethnic diversity of the region had disappeared.

Coal Miners and Unions

Coal miners often worked long hours for low wages. Living in isolated company towns allowed coal operators to exercise considerable control over their workers. Coal companies had their own law enforcement systems, often independent of local authorities.[29] Religion, education, alcohol consumption, and entertainment were regularly supervised by company managers. Coal companies controlled nearly every aspect of local life in Big Sandy Valley coal towns. Businessmen, professionals, and managers were dependent on the coal companies. Coal miners lived in company houses and were tenants of the coal companies. Coal operators appointed and paid sheriffs and deputies, influenced newspapers, and hand-picked candidates for public offices. Several trade unions tried to organize coal miners in the 1870s and 1880s with little success.[30] In 1890, at Columbus, Ohio, two of the largest organizations, a branch of the Knights of Labor Trade Assembly and the National Progressive Union of Miners and Mine Laborers, merged to form a new national union to organize coal miners: the United Mine Workers of America (UMWA). By the early 1900s the UMWA began organizing efforts in the southern Appalachian coalfields.

The outbreak of the First World War brought changes to the Big Sandy Valley and people around the nation began to realize

A coal operator and his sheriff. Courtesy of Appalachian Photographic Archives, Alice Lloyd College.

the importance of coal. The great demand for coal in industry led to severe shortages in the supply for home heating, causing prices to soar. In some cities there were coal riots as desperate people fought over limited supplies. As war production dramatically increased, President Woodrow Wilson's federal fuel administration called for the mining of record amounts of coal. Coal companies responded by expanding their mining operations, building new company towns, and reaping great profits from the booming industrial economy. The Wilson administration enforced mining laws, wage agreements, and prices and allowed miners to organize, but not strike, during the war. The war economy boosted wages, but to maintain peak production the miners worked long hours and had few days off. More mining resulted in more mining accidents, and in West Virginia alone 404 men died in mining accidents in 1918. With the Armistice of November 1918, production slowed and the boom economy collapsed. Relations soured between the UMWA and the Wilson administration and the union planned a major strike in America's coalfields.[31]

Union Battles in West Virginia

By 1900 the UMWA had been successful in organizing the coalfields of Pennsylvania, Ohio, and the Midwest but had made little progress in southern West Virginia. In April 1912 there was a strike in the Paint Creek–Cabin Creek area where thousands of miners rose up to challenge the coal operators of Kanawha and Fayette Counties. The companies hired Eureka, Pinkerton, and Baldwin-Felts detectives to patrol their properties. Mine guards erected barricades, installed machine guns, and built fortresses to protect company properties and attack striking miners. Company guards and detectives ruthlessly evicted miners from their homes, isolated and attacked miners' tent colonies, and beat union organizers and sympathizers. The fighting escalated and miners ambushed company guards, sabotaged mining equipment, and attacked trains. The governor of West Virginia finally sent in the state militia to bring an end to the fighting. From the ranks of the UMWA and radical organizations that had joined the strike, a new group of labor leaders emerged, including Bill Blizzard, William Petry, and Fred Mooney. The Paint Creek–Cabin Creek strike sent a chill through West Virginia coal operators. To combat the unions, they formed coal operators' associations throughout the state and placed informers among the workers. The Tug Valley began to look like a foreign country where company guards and local officials kept an eye on all who entered and left the region. Anyone suspected of union activities was promptly fired and evicted from company property.

The West Virginia Mine War of 1920–21

After the First World War, West Virginia entered a new cycle of violence that culminated in the West Virginia Mine War of 1920–21. On 4 September 1919 several hundred miners gathered at Lens Creek near Marmet in a remote part of Kanawha County. They had heard stories of brutal attacks on union men and their families in nearby coal towns. By the next afternoon, there were five thousand armed and angry miners preparing for a march on the town of Logan, thirty-five miles away. UMWA District 17 President Frank Keeney feared the situation was becoming dangerous and worked with Gov. John J. Cornwell to bring the crowd under control. After hours of negotiation, the march was called off, the crowd dispersed, and a formal investigation followed. The first attempted march by an angry crowd of armed miners on Logan had been prevented.[32]

A nationwide coal strike made matters worse, however. On 20 January 1920, UMWA President John L. Lewis announced plans to organize coal miners in southern West Virginia. Frank Keeney launched a membership drive in Mingo County, where the union received an enthusiastic response. Many coal operators were caught off guard, but they quickly made plans to halt the advance of the UMWA. During the southern West Virginia membership drive, the national strike was settled. Union coal miners received a 27 percent pay raise and President Wilson agreed to the appointment of a commission to thoroughly investigate living and working conditions in the coal industry. This encouraged the union, and in March 1920 they stepped up their membership drives in Mingo, Logan, and McDowell Counties. Keeney, accompanied by Billy Blizzard, the president of the local subdistrict, spoke in Matewan, drawing hundreds of miners from nearby coal towns. Mary Harris "Mother" Jones, the ninety-year-old veteran of many labor disputes, revered as a saint among miners, fired up the crowd. Hundreds of miners paraded through Williamson and held a rally on the Mingo County Courthouse lawn. Coal operators refused to negotiate with Keeney; instead, they brought in more Baldwin-Felts guards. This set the stage for the Big Sandy Valley's most explosive and violent period of labor unrest. During the next two years, the state was the scene of assassinations, assaults, and armed insurrections. The West Virginia Mine War had begun.[33]

Tug Valley coal operators quickly responded. Every miner at Matewan's Stone Coal Company was fired after joining the union. At Red Jacket, five hundred men were dismissed for attending union-sponsored events. Baldwin-Felts detectives carried out the evictions, and desperate homeless miners and their families huddled in tent camps in remote hollows along the Tug Fork. In Matewan, detectives and company guards clashed with local authorities, including Albert Sidney "Sid" Hatfield.[34] Hatfield was born in 1893 in the Blackberry Creek area along the Kentucky–West Virginia border. He was a small, thin man about five feet six inches tall. He entered the mines at a young age and was promoted to a position as a skilled laborer outside of the mines, moving coal cars to the tipple. Hatfield made friends easily and was well respected despite his reputation for drinking, gambling, and fighting. After a few years he left the mines and was appointed Matewan's first chief of police, working closely with the town's mayor, Cabell C. Testerman. Testerman was a chubby, pleasant man who liked to wear suits with suspenders and bow ties. He operated a jewelry store in town where people could also buy tobacco products, gadgets, and instruments. Inside the store was a soda fountain that was popular with local miners and their families. Testerman and his young wife, Jessie, were well liked in Matewan and were known to sympathize with the miners. In the spring of 1920, Hatfield and Testerman often found themselves caught in the conflict between coal company guards and Matewan residents, many of whom were joining the union. They were particularly upset with Baldwin-Felts agents, who were forcibly evicting miners from their homes.[35]

On 19 May 1920 thirteen Baldwin-Felts detectives arrived in Matewan, including two of the Felts brothers who ran the agency. The detectives spent the afternoon working for the Stone Mountain Coal Company, removing miners' families from their homes on Mate Creek. Angry miners called Hatfield and Testerman to the scene. An argument developed between Albert Felts, Hatfield, and the mayor, but the evictions continued. Hatfield and Testerman returned to town and phoned county officials in Williamson. They found out that the agents had not been authorized to carry out the evictions. After the call, Hatfield was reported to have said, "We'll kill the goddamn sons of bitches before they ever get out of Matewan." Later that afternoon, the Baldwin-Felts men returned to Matewan

BORDERLAND

PUT YOUR NAME ON THE

OR ELSE

BOLSHEVISM	**UNIONISM**
SOCIALISM	**TERRORISM**
ANARCHISM	**BULLETS**

Ask any one of thousands who have refused to sign Union Contracts, or those who have been beneath the heel and domination of "Unionism."

Not the half of the activities, "direct action," intimidation, etc., of the Union and its sympathizers at Borderland has been told.

Unionism is all right, theoretically, perhaps; so is Socialism, theoretically; but, practically, it doesn't work; and "work" here has a double meaning.

After signing a Union Agreement the real trouble, petty annoyances, unfairness born of supposed power, begin.

"Heavy, heavy hangs over your head," all the time.

If one signs an Agreement for, say, one year, and then does not want to renew it, he is branded as "non-union," "rat," "scab," "unfair"; is boycotted and suffers from all the annoyances Union Agents and their sympathizers can inflict.

Which is quite different from what happens between business people when they do not sign or renew a contract.

Power takes its morality from the results of its use. If the power of trades unionism means that a man cannot work without the approval of a union, no matter what its motto for the future, no matter what its record for the past, no matter who its champions, trade unionism cannot survive in the United States.

—LAW AND LABOR.

"Two-Gun" Sid Hatfield. West Virginia and Regional History Collection, West Virginia University Libraries.

and waited for the five o'clock train to Bluefield. Hatfield approached the men with warrants for their arrest, signed by Testerman. As the men argued, Mayor Testerman was called to the scene, and a large crowd of angry, armed town residents gathered around the men.

No one is sure how the shooting started. Reliable sources say that Albert Felts fired first, hitting Mayor Testerman. Hatfield drew his gun and shot Felts dead. Sid then drew his other gun and fired away. The miners opened fire on the detectives. While returning fire, the detectives tried to disappear into the crowd. The angry crowd chased the men through Matewan, gunning them down in the streets. Troy Higgins, a former Virginia police chief, was killed as he fled the scene. C. B. Cunningham was shot so many times that he was nearly decapitated. Two other detectives, A. J. Booher and E. O. Powell, were killed as they ran through the streets. J. W. Ferguson was wounded but was taken into a nearby home. The angry crowd dragged him out and killed him.

Some of the Baldwin-Felts men escaped. Oscar Bennett was lucky; he had left the scene just as Hatfield and Felts began arguing, slipping unnoticed into one of the passenger cars. Two other brothers, Walter and Tim Anderson, hid during the shooting and were able to board the train to Bluefield. John McDowell, another former policeman, was chased to the river, but he made it across and fled into the Kentucky woods. Two miners were killed during the gunfight. Bob Mullins, who had just been fired for joining the union, was killed by Booher. Tot Tinsley was also found dead after the fight, but it is unclear who shot him. When it was finally over, seven detectives (including both Felts brothers), two miners, and the mayor were dead. Five others were wounded in the fight and six detectives escaped. Hatfield earned the nickname "two-gun Sid," the hero of the Matewan Massacre.[36]

Ten days later, Sid Hatfield married Jessie Testerman, the mayor's young widow. Despite rumors that he may have intentionally brought Testerman into the fatal conflict, Sid Hatfield remained extremely popular in Mingo County. Local authorities were less impressed. They charged Hatfield and twenty-two others with the murder of one man, Albert Felts. The trial, which took place in February and March of 1921, was one of the most dramatic in the history of the state of West Virginia. Federal troops guarded the courthouse and large crowds gathered at the scene. The state had a weak case and offered immunity to several defendants, hoping to gather evidence against Sid. Some of the most interesting testimony was given by one witness, Charles E. Lively, who turned out to be a double agent. Lively ran a popular local restaurant, was involved in many union activities, and had become a friend and confidant of Hatfield. During the trial it was revealed that he worked for Baldwin-Felts. He testified that on several occasions he had heard Sid bragging that he had shot Felts dead. Lively had actually been conducting his own investigation of the massacre and had relayed much of the information directly to Thomas Felts. The final arguments were presented on Saturday, 18 March 1921. On Monday the jury returned a verdict of not guilty, acquitting Sid Hatfield and the remaining sixteen defendants. Sid and Jessie returned to Matewan and received a hero's welcome. Almost the entire town met them at the depot. The union made a silent movie called *Smilin' Sid* that became one of the most popular films in the coal camps of West Virginia.[37]

In Matewan, a new burst of violence followed the trial. Beginning on 12 May 1921, almost one year after the massacre, shootouts raged between union men and company guards, resulting in several deaths. Trains raced through the valley to avoid the gunshots, and the businesses of Matewan shut their doors as snipers fired throughout the town. Only through the negotiations of a local physician was the fighting finally brought to an end. The three-day battle alarmed area residents. President Harding and Gov. Ephraim Morgan were called on to declare martial law. In Williamson, a citizens militia was organized to defend the town. On 13 June there were shots fired at state police, and the militia was called out to attack a tent colony of evicted miners at Lick Creek. They opened fire with machine guns, ran women and children into the woods, and arrested more than forty men. The attack on Lick Creek received widespread attention and prompted a U.S. Senate investigation.[38] Sid Hatfield was summoned to Washington, D.C., to testify before the Senate Committee on Labor and Education, where he helped draw national attention to the

Mother Jones (center) with labor union organizers in West Virginia. Eastern Regional Coal Archives.

plight of miners along the Tug Fork. Following his testimony, Hatfield reported to the McDowell county seat of Welch to face other charges. When he arrived on the steps of the courthouse with his close friend Ed Chambers on 20 August 1921, he was met by a group of Baldwin-Felts men led by double agent Charles Lively. Lively and the guards shot Sid Hatfield and Ed Chambers dead on the courthouse steps. Thousands attended their funeral in Matewan.[39]

The death of Sid Hatfield sparked a new series of confrontations and the final phase of the West Virginia Mine War. In mid-August 1921, at Lens Creek, about ten miles from Charleston, a new army of miners was formed, and within a few days thousands of armed volunteers from Boone, McDowell, Raleigh, Fayette, and Wyoming Counties joined their ranks. The miners planned to march through Logan County, ridding the coal towns of Baldwin-Felts men and local sheriffs and deputies who worked for the coal operators.

Their main target was Logan County's notorious sheriff Don Chafin, a feared and hated man who ran the county like a feudal lord. He had worked his way up through the corrupt political system and was a master at political manipulation and intimidation. Chafin, born in 1887, attended Marshall College in Huntington and returned to Logan to enter politics. In 1912 he won election as sheriff and controlled the office for the next twenty years. When he was not serving as sheriff, he was able to hand-pick successors and keep tight control over the county's many deputies. Although many coal operators hired private detective companies to oversee their mines and coal towns, in Logan County, Chafin offered them a better deal. His army of deputies harassed union organizers, monitored highway and railroad traffic, and brutally dealt with those who resisted. As law enforcement officers, they had many advantages over typical company guards. Chafin's force was so effective that he contracted his services directly to the coal companies and began to build a personal fortune of more than $350,000 by the early 1920s. Chafin's power spread far beyond Logan County, with his deputies occasionally attacking state officials and reporters.[40]

As the miners' army advanced, Chafin worked with Logan County coal operators to prepare a defense. He recruited thousands of volunteers, deputized hundreds of men, and had hundreds of state police at his disposal. Coal operators even arranged for the use of several private airplanes. But as Logan prepared for the siege, Chafin became desperate for more recruits.

He emptied the jails and forced inmates to fight with his men. The miners, who hated Chafin, his deputies, and their tactics, planned to capture and execute Don Chafin after seizing the county seat of Logan.

The ranks of the miners' army grew to almost twenty thousand. Countless tens of thousands of area residents aided and assisted the army as it began its march. In addition, the miners' army had doctors, nurses, security patrols, and cooks. An advance force cut through the area south of Charleston and cleared the way by removing telephone wires, telegraph lines, and Baldwin-Felts men. They hijacked trains and moved men and supplies into settlements along the Little Coal River and Spruce Creek. Moving south through Sharples, the miners' army finally arrived at the base of Blair Mountain and began their ascent. Small bands of armed miners approached the outskirts of Logan. Chafin's aircraft bombed selected targets. There was a pitched battle, with the line of fire reaching from Blair Mountain to Crooked Creek, a distance of twelve miles. Chafin's army captured miners as they moved more boldly toward the town. By Friday evening they were rounding up miners and holding them in the Logan jail. The governor organized state militia units, but federal forces were soon brought into the conflict. The U.S. Army organized its Eighty-eighth Air Squadron for the West Virginia operation. They were equipped to carry bombs, machine guns, and supplies. Several Martin bombers were also mobilized. As they headed for southern West Virginia, bad weather and problems maneuvering through the mountain passes took their toll. Only fourteen of the twenty-one planes ordered to the state made it. A crash on Saturday, 3 September, killed four crewmen. Meanwhile, federal troops began arriving from Camp Dix outfitted with rifles and machine guns. Trainloads of support troops made their way from Indiana and Ohio. The plan called for about two thousand troops to be used in the operation. As they arrived at Sharples, the "capital" of the rebellion, they began to round up hundreds of armed men and free trains that had been commandeered by the miners.[41]

The deployment of troops and aircraft to West Virginia brought the national press, and small bands of reporters converged on the area. One group, led by veteran *New York Tribune* war correspondent Boyden Sparks, wandered into the mountains and found themselves under attack by state police. Sparks was wounded, and he and the other reporters were promptly taken back to Logan. Neither side respected the rights of the press, and many correspondents found themselves harassed, threatened, and prevented from filing their stories. Despite the problems, the battle of Blair Mountain drew unprecedented national attention.

Federal troops moved through the hills and hollows capturing hundreds of miners. Most were disarmed, shipped out of the area on rail or returned home without incident. The army set up a headquarters at Madison and kept troops in place at Sharples, Jeffrey, and Clothier. The anticipated fight never quite took place. Gunfire was limited and most miners surrendered without a struggle. The miners' army, which had appeared so vast and threatening, was dismantled in a matter of days. Despite the reconnaissance flights, thousands of troops, and the constant gunfire heard throughout the mountains, there were few casualties. On Sunday, troops moved into Blair and began to scour the countryside for bodies, but few were found. In the entire operation only twelve miners and four men among the citizens militias and police were killed. By Sunday afternoon, the mission was over and Logan returned to normal. Many of the militia and volunteers returned home and were greeted as heroes. They had prevented Logan County from being taken over by a mob. Within days the U.S. Army was getting ready to leave. The battle of Blair Mountain and the West Virginia mine war had ended.

After the troops' departure, local officials regained control of Mingo and Logan Counties. More than twelve hundred indictments were handed

U.S. Army supply train with troops arriving at Logan, September 1921, during the West Virginia Mine War. Chesapeake & Ohio Historical Society.

down for incidents relating to the two-year labor struggle. The UMWA called off the strike. Membership in southern West Virginia plummeted. For the rest of the decade, things remained unchanged in southern West Virginia. The same sheriffs and police patrolled Mingo and Logan Counties; coal operators continued to use Baldwin-Felts agents as mine guards, and unions gave up on organizing mines in Mingo, Logan, and McDowell Counties.[42]

Instead, the union concentrated its efforts in the northern coalfields of the state. Although the violence and the mine war continued throughout the state for years, union membership dropped sharply and continued to decline for the rest of the decade. From its peak of about fifty thousand members, the union could claim less than six hundred in West Virginia by 1929. Although there were more than one thousand indictments in the months following the Battle of Blair Mountain, including 324 for murder, few people were ever convicted. Local governments abandoned many of the prosecutions and most of the indictments were dismissed.[43]

Although West Virginia experienced great conflicts between coal operators and union men during the 1910s and the 1920s, there was little impact in east Kentucky and southwest Virginia. The union's efforts affected miners directly across the Tug Fork but did not reach beyond the mountains into Pike County coal towns. Organizers were sent into the two bordering states, but West Virginia remained the focus of their efforts. The collapse of the southern West Virginia movement greatly weakened efforts in the rest of the Big Sandy River Valley.[44]

The 1920s were a difficult time for the coal industry. Coal companies had overexpanded and overproduced. The price of coal dropped sharply and coal markets were glutted. Coal companies continued to lay off workers as production slowed. Lewis and the UMWA accepted agreements that would hold wages at their current rates but would not protect jobs. To make matters worse, coal operators introduced record numbers of machines in the 1920s, reducing the need for laborers. During the remainder of the decade, the number of coal miners in the United States dropped from more than 700,000 to less than 575,000. As competition for mining jobs increased, coal companies reduced wages. Desperate for work, miners signed "yellow dog" contracts that made working as a non-union miner a condition for employment. For the UMWA, the losses were staggering. With little bargaining power against the coal companies and a series of Republican presidents and congresses hostile to labor unions, national membership dropped from more than 500,000 in 1921 to less than 80,000 by 1928.[45]

Meanwhile, the U.S. Coal Commission, created during the Wilson administration, made a detailed study of the coal industry. It surveyed living conditions in coal towns, examined health and safety issues, and made recommendations for further regulation of the industry. Unfortunately, the study was undertaken at a time when the coal industry was already suffering. By the time its multivolume report was published in 1925, the federal government had eased many regulations, making its recommendations weak. The 1920s saw an unprecedented number of mergers and consolidations in major industries. Large coal companies now had the resources, technology, public relations power, and political influence to effectively crush any challenge from labor.

A slump in the coal economy created other problems. Tensions often developed between native white mountaineers, African Americans, and immigrant workers. As a result, the Ku Klux Klan spread throughout the Big Sandy Valley during the 1920s. Paintsville hosted a mass Klan rally on Labor Day, 1924, that attracted five thousand people. The Klan held similar events in east Kentucky and some West Virginia communities during the decade. The Klan members were largely dedicated to enforcing traditional Protestant religious values with regard to alcohol, gambling, and sexual morality, but many were also anti-immigrant. In addition, the national fear of foreigners led to quotas and restrictions, and few immigrants came to the Big Sandy Valley after 1920. In West Virginia, the Klan was less successful, and the UMWA may have helped keep their influence minimal.[46]

The miner's situation was worse after the stock market crash of 1929. Some labor leaders became militant and demanded more government intervention in the economy. Lewis used the Great Depression as an opportunity to revive the UMWA. One major reason that labor emerged so successful after 1932 was the election of Franklin D. Roosevelt to the presidency and a series of New Deal acts that legitimized and aided labor unions. A series of strikes took place in 1932–33 in Pennsylvania, West Virginia, and Kentucky. The UMWA took advantage of the opportunity and negotiated the Appalachian Agreement, which became effective in October 1933. It provided for an eight-hour workday and a five-day workweek. The payment of scrip was prohibited, and miners were no longer required to live in company housing or shop at company stores. Some coal companies used tactics that had been successful in the 1920s to block union's efforts, and Kentucky's Harlan County became the place of a prolonged civil war between union and company men. By the mid-1930s, the county was nationally known as "bloody Harlan County" and attracted journalists, organizers, and political activists from around the nation. The creation of a National Labor Relations Board protected the workers' right of collective bargaining, and a bill introduced by New York Sen. Robert Wagner, commonly known as the Wagner Act, empowered labor unions. The movement was further aided by the creation of a congressional Civil Liberties Committee that investigated the use of company guards, detective agencies, weapons, and espionage tactics by leading coal companies. Although some areas experienced violence and conflict similar to that of the 1910s, most parts of the Big Sandy River Valley made a peaceful transition to unionism. In fact, many companies hoped that the settling of wage disputes and some government regulation might stabilize an erratic industry and make way for modernization. Most of the resistance came from small coal companies that would no longer be able to compete by using poorly paid nonunion workers. Lewis made great gains in the Big Sandy River Valley coalfields by the late 1930s, including control of "captive" mines, mines that were operated by other industries. Major coal towns operated by the steel industry such as Lynch, Kentucky, and Gary, West Virginia, were soon union towns.[47] Although there were revisions made in later contracts, Lewis had gained control of the Appalachian coalfields, and by the 1940s almost the entire bituminous industry was organized. The policies and codes established helped stabilize coal prices and by the late 1930s they began to rise. Although this was a success for labor unions, coal miners paid a price. Companies soon focused on modernization and mechanization as a way to increase production and reduce the number of miners needed.[48]

The Second World War and Decline

The Second World War temporarily reinvigorated employment opportunities in the coalfields, but it also caused people to leave the region. The expansion of war-related industries in Ohio, Indiana, and Michigan attracted workers from throughout Appalachia. After the war, less coal was used for generating electrical power, and oil and natural gas replaced coal for heating homes. The change from steam- to diesel-powered locomotives further reduced the demand for coal. During the 1950s, the

UMWA lost power and influence. Mechanization reduced the number of jobs in the mines, increasing workers' migration north and reducing union membership. Truck mining also changed the nature of coal mining by the 1940s. Roads opened new areas to mining, and trucks transported the coal to the railroads. This opened hundreds of small mines in the Big Sandy Valley. Small operators were subject to less regulation and could hire nonunion miners at lower wages. They could operate mines without building elaborate transportation systems or company towns or employing large numbers of workers.

The declining demand for coal, the need for fewer workers, the development of truck mining, and the transfer of worker loyalty from the employer to the union convinced coal executives to abandon company towns. In the fifteen years following the end of the Second World War, most coal corporations sold the houses in their towns to the residents. Other property was sold or leased to smaller companies willing to extract the remaining coal reserves.[49] The coal industry would remain a vital part of the economy of the Big Sandy Valley for the remainder of the twentieth century, but coal companies would no longer have the same control over the lives of their workers or guarantee the same standard of living. Only a few thriving coal towns such as Lynch and Wheelwright, Kentucky, and Gary, West Virginia, most under control of the steel industry, continued to be company towns. The UMWA accepted the fewer jobs in the mines in exchange for higher wages and benefits for union workers.[50]

UMWA President John Lewis hoped that the union could provide some of the same services that coal companies had offered during the era of the company towns. In the mid-1950s the union used monies from its Welfare and Retirement Fund to build ten regional hospitals in southern Appalachia to serve coal miners and their families. The Miners Memorial Hospital Association was created to run this health care system and to provide extended services to remote mountain communities. The system worked well for a few years, but by the early 1960s it faced severe financial problems and collapsed in 1963. The ten regional hospitals were purchased by a new nonprofit organization, Appalachian Regional Hospitals, which received generous private, state, and federal funds to continue operating. Today, these hospitals still operate in the Big Sandy region.[51]

The economic stagnation of the 1950s and 1960s gained Appalachia, and especially east Kentucky, a negative image. Writers like Harry M. Caudill of Whitesburg, Kentucky, drew a national audience as they described the poverty and desperation of the region. Reporters and social workers converged on the region, describing the squalid conditions and simple, isolated people of the mountains who were now abandoned and destitute. These simplistic portraits created great sympathy for southern Appalachia, which led directly to a series of programs designed to stimulate economic progress in the region.[52]

John F. Kennedy set a new agenda by visiting depressed coal towns in West Virginia during his presidential campaign in 1960. He promised to help revitalize the area and provide basic social services to remote mining communities. In Kentucky, these new efforts were led by Gov. Bert T. Combs. Roads were improved, schools and colleges were expanded, and the mountain parkway was constructed, linking the Big Sandy region with Lexington.[53] Kentucky Congressman Carl D. Perkins worked closely with President Kennedy and President Johnson to design federal programs to aid the poor. While serving on the House Education and Labor Committee during the 1960s, Perkins helped initiate Johnson's War on Poverty and brought further attention to the region. President Johnson journeyed to the Big Sandy Valley in 1964 promoting his antipoverty programs. West Virginia also instituted new reform programs in the 1960s and 1970s. Under Republican Gov. Arch Moore the state college system was expanded, new mining regulations were drafted, and pneumoconiosis, or black lung, was classified as a disease. Economic development programs also focused on diversifying the economy of the state. A State Department of Highways was created to help achieve this goal. In the 1970s and 1980s, Governor and Senator John "Jay" Rockefeller broadened miners' rights and challenged strip mining companies. During the Carter administration he headed a commission that studied the coal industry.[54]

In 1965 the Appalachian Regional Commission (ARC) was established as a federal agency to help develop the Appalachian areas of thirteen states. Much of eastern Kentucky and the entire state of West Virginia was included in this region, which spanned 397 counties and covered twenty million people. In its more than thirty-year history, the ARC has brought more than $7 billion into the region for highway construction, health care, utilities, vocational training, research, and technical assistance. During the 1960s and early 1970s, the ARC and War on Poverty programs brought hundreds of volunteers into the region. Local program for high school retention, childcare, childhood development, recreation, and housing were initiated. Community action, preschool education, and job training programs were also developed.[55] Despite the fact that a substantial social service network was developed that provided impoverished Big Sandy Valley residents with food, shelter, medical care, education, and training, the out-migration continued. Few new employment opportunities were created, and those that were created seldom paid as much as a good job in the mines. Many residents were caught in a cycle of relying on public assistance that, in many cases, has persisted for generations.

Recent Changes in the Big Sandy River Valley

The Arab oil embargoes of the 1970s sparked a second, but brief, coal boom. Increased mining activity stimulated employment and created instant millionaires. It also temporarily reversed the flow of migrants, with many natives returning from the North to take high paying jobs in the coalfields.[56] The crisis also helped bring about a minor change in the workforce: a small number of coal companies hired women to work in the mines. By the early 1980s, there were about thirty-five hundred female coal miners in the country.[57] The 1970s also saw a renewed interest in the plight of miners. The 1969 Federal Coal Mine Health and Safety Act opened the way for miners to receive black lung benefits. By the 1970s, much of the coal extracted from the Big Sandy Valley was taken from surface or strip mines. In the process, mountaintops were leveled, vegetation was removed, and rockslides became increasingly common. The battle against strip-mining began prior to the coal boom of the 1970s, but the expansion of that practice during the energy crisis led to an outcry by environmentalists. The Kentucky Department of Natural Resources began to enforce the state's strip-mine laws in 1974.[58] Congress moved to strengthen federal regulations in 1977. In 1988 Kentuckians for the Commonwealth helped to force the passage of an amendment to the Kentucky Constitution requiring a landowner's permission before strip-mining could take place.[59]

Strip-mining had also become more sophisticated. In the 1970s, bulldozers were used to dig ledges around mountainsides, and coal was then drilled out from the exposed seams. The latest methods use draglines to shear the surface off mountaintops and expose rock and coal seams below. Draglines can cover areas of up to two hundred feet and weigh millions of pounds. The surface is stripped away and rock layers are blasted off. The exposed coal is then shattered, loaded into trucks, and driven away. The topsoil and foliage stripped from the mountain are dumped into nearby valleys, often filling ravines and changing drainage patterns. This particular form of strip-mining has generated great controversy because it levels out large areas and changes the landscape significantly. This method also requires far fewer workers than traditional mining. In West Virginia, for example, the number of coal mining jobs has been reduced from more than 100,000 in the 1950s to less than 22,000 today. Now entire mountaintops are being removed throughout the Big Sandy region, having an impact on the environment and threatening many communities.[60]

The coal boom of the 1970s lasted a short time. During the 1980s and 1990s the coal industry remained depressed. Only in the beginning of the twenty-first century has there been an increase in coal mining because of higher oil prices. Realizing the limitations of the coal industry in the Big Sandy Valley, community leaders have more recently looked for other means of economic development. Light manufacturing and service centers have come to some areas of the Big Sandy region. Major improvements of U.S. 119 between Charleston, West Virginia, and Pikeville, Kentucky, will no doubt attract further commercial activities.

Tourism is also emerging as an important part of the local economy. Recreational lakes, a summer theater, and the Kentucky Opry attract thousands of visitors to the area each year. Efforts are under way to preserve the area's heritage in art, music, and folkways. The Mountain Home Place at Paintsville Lake, Kentucky, and the new Mountain Arts Center in Prestonsburg will play a major role in accomplishing this task. The Matewan Development Center in Matewan, West Virginia, has been active in promoting the sites of the Hatfield-McCoy feud and preserving the coal mining heritage of the region. The renovation of Benham and Lynch, in Harlan County, Kentucky, also promises to attract tourists as well. In 1998 an eleven-county region of West Virginia was designated as a National Coal Heritage Area, and parts of U.S. 52 and WV 16 are being developed as a Coal Heritage Trail. The restoration of towns and the creation of new tourist services might inspire other communities to look to their coal mining past as a means of future economic development in this century.

3

Coal, Coal Mining, Coal Towns, and Structures

Coal

Coal is a rock that can be ignited and burned. It is used as a fuel to produce high temperatures. Today's coal was formed between four hundred million and one million years ago from the fossilized remains of plants, which is why it is called a fossil fuel. As plants died, they formed a thick layer on the swamp floor, and over the years these layers turned into peat. The peat became buried under layers of sand, clay, or silt, which gradually compressed the peat into coal. Under pressure, the layers of sediment became shale and sandstone. Shale, sandstone, and coal are classified as sedimentary rocks. There are three major types of coal: lignite, anthracite, and bituminous. As peat is pressurized, it develops into a dark brown form of coal called lignite. As the pressure increases over time, lignite turns into a more compressed form, bituminous coal. Extreme pressure turns bituminous coal into anthracite coal. Anthracite coal is not necessarily older than bituminous; it is simply formed by additional heat and pressure. Bituminous coal is often referred to as soft coal, and anthracite, as hard coal.[1]

Anthracite coal is dense, heavy, and black and has the highest carbon content. When ignited, it burns with a blue flame and produces little soot. Because it has little moisture, it produces high temperatures and is said to have a high heating value. Anthracite coal is the least plentiful coal in the United States. It is difficult to ignite and burns slowly. Nearly all of the nation's anthracite coal is found in the Northeast, especially in Pennsylvania. Bituminous coal is not as compressed as anthracite coal and has a lower carbon content. It is lighter and has a dull black sheen. Bituminous coal is the most common coal in the United States and has the widest range of uses. It ignites more easily than anthracite coal and burns quickly with a smoky yellow flame. Bituminous coal is the only coal that can be used to produce coke, a fuel used in blast furnaces to smelt metals. All of the coal mined in the Big Sandy River region is bituminous coal. Lignite is a high-moisture coal with a lower heating value than anthracite or bituminous coal and fewer practical applications. It is found in the western United States and Canada.

When coal is burned it releases sulfur, which combines with oxygen to create sulfur dioxide, a very poisonous gas. Low-sulfur coal can be burned in large quantities and produces little sulfur dioxide. Medium- and high-sulfur coals can be dangerous. When burned, they can release large amounts of sulfur dioxide into the air, creating a serious air pollution problem. Today, the U.S. Department of Energy (DOE) classifies the sulfur content of coal. Lignite has a low sulfur content, and anthracite coal has a high sulfur content. Bituminous coals range from medium to high sulfur content. The amount of volatile matter determines how much smoke is produced when coal is burned. Coal also produces ash when burned, and fly ash can be released into the atmosphere. During the twentieth century, smoke and ash from coal were a great problem in urban-industrial areas. The Clean Air Act of 1970 put limitations on the amount of pollutants that can be produced by coal, and the Clean Air Act of 1990 tightened these restrictions. New procedures for cleaning coal have reduced pollution as well.

Coal is found in seams that run through layers of the earth and are sometimes exposed where mountains have formed. Seams consist of a common type of coal and are found throughout a region known as a coalfield. Most of the bituminous coals in the Big Sandy region are low sulfur, low ash, low volatile coal. Some have such low amounts of volatile materials, such as those in southern West Virginia's Pocahontas coalfields, that they are known as smokeless coals. Different varieties of coal are often marketed by companies under brand names, such as Cavalier Coal from the Elkhorn seams at Jenkins.[2]

Coal is used in many products, such as dyes, drugs, and fertilizers, and is used to heat buildings, but it has two main uses in the United States today. The most important use is the production of electricity, which consumes more than three-fourths of the coal mined. Burning coal heats water and produces pressurized steam that drives turbines. The turbines drive generators that produce electricity. About 70 percent of the country's electricity is produced by steam turbines, and most of the steam turbines use coal as a fuel. The second most important use of coal is in making metallurgical fuel coke. Coke is produced by baking bituminous coal in airtight ovens at temperatures up to two thousand degrees Fahrenheit. Only certain types of low-sulfur, bituminous coal can be used as coking coal. The baking process separates volatile gases from minerals and results in a lighter, foamlike substance that is almost pure carbon. Coke is the coal industry's most important product and is used primarily in smelting iron ore. More than 90 percent of the coke made in the United States is used to make iron and steel.[3]

Coal continues to be the main source of electrical power and is vital to manufacturing. The amount of coal mined in the United States every year has grown, and exports, as well as domestic use, continue to increase. In fact, the total amount of coal mined has more than quadrupled since 1900 and has more than doubled since the 1960s. Although coal is no longer the main overall source of energy (petroleum and natural gas have replaced it in many uses), it continues to be used in record amounts. In the 1990s, the United States mined more than one billion tons of coal annually. In 1999 Kentucky and West Virginia produced more than 156,000,000 and 169,000,000 tons of coal respectively.[4]

Coal Mining

Although no one knows where or when man first found out that coal could be burned as a source of heat, the earliest place where coal was commonly used was China. Around 300 A.D. the Chinese were mining surface deposits of coal. By the 1270s, when Venetian explorer Marco Polo traveled through China, he noted that it was used to heat buildings and smelt metals. In Europe the common use of coal developed much later. In the fourteenth century the depletion of northern England's forests led to a shortage of wood and a market developed for substitute fuels. Serfs were used to dig surface deposits of coal, and by the 1500s vertical shafts were being dug to reach coal seams. English coal mines became known as collieries, a name that was later applied in some coal mining areas of the United States.[5]

Access to coal was limited. When digging shafts into the ground, workers quickly reached the water table. Primitive pumps were designed to remove water from the mines but had limited capacities. A steam engine designed by Thomas Newcomen in 1712 pumped water out of mines. During the 1770s James Watt introduced another design that could significantly increase the amount of water drawn. During the late eighteenth century Watt's steam engine was applied to factory machinery. By 1800 Great Britain's Industrial Revolution was well under way and ironmasters were using coke to smelt iron ores. Iron furnaces and ironworkers' shops increasingly used coal. In the United States the change to coal was a bit slower, but many iron furnaces were using it by the 1850s.[6] America's industrial development lagged behind Great Britain, but the rapid growth of railroads, new demand for steel machinery, and the use of coal as a fuel dramatically increased by the time of the Civil War. Eastern Pennsylvania and parts of Maryland became the main sources.

There was a big increase in the use of coal after the Civil War. In the 1840s, railroads needed a metal stronger than iron. The rich iron-ore deposits of the upper Great Lakes and the anthracite coals of Pennsylvania were soon combined to produce steel. Steel was stronger than iron and necessary to build longer bridges, stronger railroad track, and machine parts. In the 1850s Englishman Henry Bessemer developed a new process for the mass production of steel. Similar innovations were made by American William Kelly. Bessemer and Kelly discovered that blasts of air into molten iron could burn off more impurities and result in a purer, stronger steel. This process revolutionized American industry and developed great steel districts in Pennsylvania, Alabama, and Ohio, where large coal deposits were available. Railroad locomotives also began using coal instead of wood as the standard fuel for driving steam engines. Coal was also commonly used for home heating, replacing wood in the cities by the 1890s.[7]

Although manufacturing steel, powering steam engines, and heating buildings were important functions of coal in the late nineteenth century, they were soon overshadowed by the need for coal to generate electricity. Electricity powered trolley cars, appliances, and industrial machinery. Residential lighting created more demand by 1900, and cities built power stations to supply electricity to industries, businesses, and homes. At a power station, coal was used with water to make steam, which drove generators that produced electricity. Alternating current allowed high-voltage electricity to be transmitted by wire. The need for water and coal placed power stations along rivers, in harbors, and railroad lines, where both resources were available. Power stations stockpiled huge amounts of coal to compensate for shortages. They became a common sight in cities with their massive smokestacks and mountains of coal piled on one side and ash heaped on the other. Bituminous coal was cheaper than anthracite and became the preferred stock.[8]

British firms invested heavily in American companies, and much of the early coal industry benefited from British capital, technical expertise, and labor. Mining methods, terminology, and labor were brought to the States. After 1870 the nation's economic growth reached new heights and coal mining expanded. The Big Sandy Valley was known as a source of coal for some time, but transportation problems limited access to coal seams and markets. Flatboats, barges, and steamships were used to float coal downstream to markets in the Ohio Valley. As railroads penetrated remote areas, coal mining became more profitable.[9]

Coal Mines

There were three basic types of underground mines in the bituminous coal industry in the Big Sandy region in the late nineteenth and early twentieth centuries. The first and most common type was the drift mine, where an opening was dug horizontally into the side of a mountain to reach a seam of coal. These mines were located above the valley floor and were the least expensive to construct and work. There was no need for elaborate transportation or ventilation systems to be built. Miners simply walked in the entrance to their work sites, and horses or mules dragged coal cars from the mine. Coal was then moved down to a tipple for sorting. As drift mines became more sophisticated, rails were laid and small locomotives transported coal cars and miners. Because coal seams were located high above a valley or hollow, coal was easily moved down the mountainside to a tipple and railroad line below. In mountainous regions such as the Big Sandy River Valley, drift mines remain common.

The second type of underground mine was the slope mine, where an inclined entrance is dug to reach a coal seam. Slope mines were useful where the coal seam was well below the surface but not at a great depth. Early slope mines were a bit more difficult than drift mines to construct and required more horsepower or manpower, depending on the angle of the slope and the length of the entrance. In the twentieth century, locomotives and conveyors made slope mining easier.

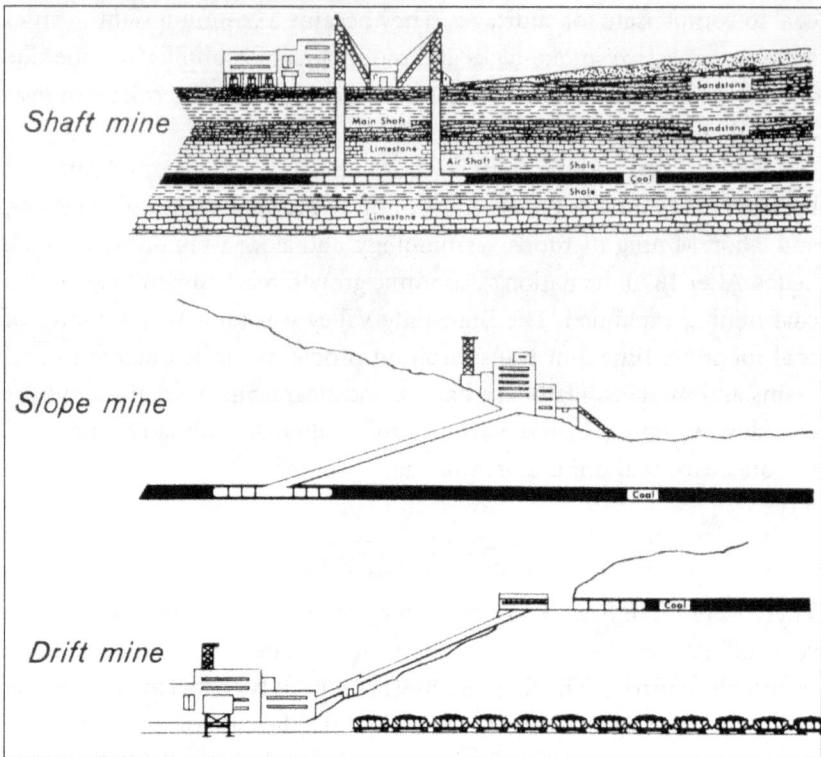

Three types of underground coal mines found in the Big Sandy Region. Courtesy of Keith Dix/Institute for Labor Studies.

The third type of mine was the shaft mine, where a vertical shaft was dug to reach a coal seam below the surface and miners and equipment were lowered to the work site with hoists or elevators. In the early twentieth century, there were few shaft mines because of the costs of the equipment and technology required to raise workers and mined coal to the surface. Today, they are much more common and many shaft mines operate in the Big Sandy River Valley.[10]

Methods of Mining

There are two basic methods for underground coal mining: room and pillar and longwalling. The room and pillar method is the most common. Tunnels, or entries, are dug into the coal seam from the entrance or a main tunnel. The tunnels are spaced apart and extend into the seam. Side

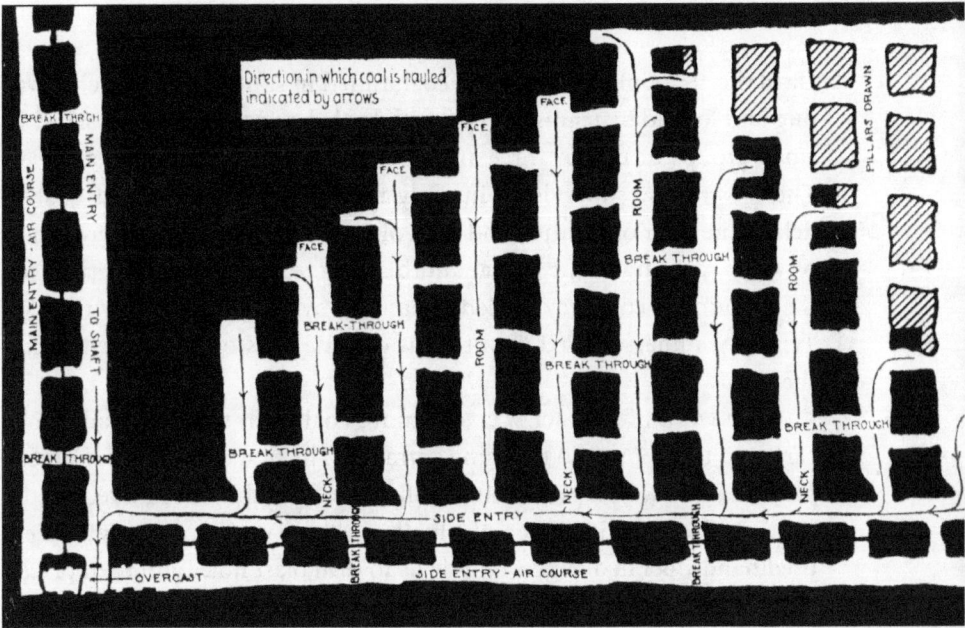

The room and pillar method of coal mining. Courtesy of Keith Dix/Institute for Labor Studies.

entries are then dug to create a grid pattern and rooms are opened to prepare the area for mining. Pillars, or columns, are left to support the roof while mining takes place. The wall at the end of the room is called a face, the site where miners remove and load the coal. Individual rooms are worked in sequence with pillars left standing for support. After all the rooms in an area are worked, the miners retreat and the pillars of coal are brought down.[11]

Longwall mining produces the most coal but requires machinery and certain structural conditions to work successfully. Instead of simply working the face at the end of a tunnel, two parallel tunnels are driven to one side, hundreds of feet apart. The area between the tunnels is mined by a large gang of workers rather than small crews. They move forward and use movable, hydraulic pillars to support the roof as they proceed. There are no coal pillars left to support the roof; it is all taken out of the mine. In the United States, longwall mining is limited by the thinness of coal seams and the weakness of mine roofs in southern Appalachia. Because of improved technology, mechanized longwalling has become much more common.[12]

The Miner's Work

Until the 1930s almost all of the coal mined in southern Appalachia was dug and loaded by teams of men with basic hand tools. They worked in small groups of two or three and bought their own tools and supplies. There were few supervisors in early mines, and miners could usually determine their own hours and pace of work. Although there were hours of setup, preparation, cutting, and blasting, miners were only paid for the amount of coal they loaded, measured by the carload or by the ton. Typically, miners loaded about eight or nine tons of coal a day during the 1920s.[13]

The work of a miner was to remove coal from the face, load it into cars, and transport it to the mine entrance. The first task was to undercut the coal seam with a pick. This usually took several hours and required a good deal of physical strength. In the early twentieth century, miners used hand-operated augers and drills to drill blast holes and then placed an explosive charge. The face was then blasted, creating a pile of coal and debris. Miners pushed their own loading cars into the rooms. After placing a token at the bottom of a mine car, coal was sorted and shoveled into the car. The token allowed a mine team to be credited for the amount of coal they loaded. In order to avoid deductions for dirty coal, they carefully sorted coal from slate and rock. Miners could only estimate how many tons of coal they had actually loaded. After the loading was completed, the cars were taken to the surface by mules or later by locomotives. In some early mines, the miners had to drag the cars to the scales.[14]

Dangers

Coal mining has always been a difficult, demanding job with great risks. Around the turn of the last century an average of fifteen hundred men lost their lives each year in American coal mines. Countless others were injured. Although the dangers have been greatly reduced in recent years, underground coal mining is still considered a hazardous occupation. Roof falls were the most common problem, killing hundreds of miners each year. Although roof falls and accidents with machinery were the biggest risks, mine explosions received the most attention. Methane gases are released during digging and blasting, and the slightest spark can set off explosions. The deeper the mines, the more risk of explosions. West Virginia has

Top: An early coal mine at Miller's Creek, Kentucky. Courtesy of the Appalachian Photographic Archives, Alice Lloyd College. Bottom: An early east Kentucky coal mine. Paul Mays Collection, Pike County Library District.

been the site of the nation's worst mine disasters. On 6 December 1907 an explosion at the Fairmont Coal Company's Monongah mine killed 361 workers. In the Big Sandy Valley, an explosion at the Pond Creek Pocahontas Coal Company's Bartley mines killed ninety-one men on 10 January 1940.[15]

Early Mechanization

As early as the 1880s, machines were being used to cut and transport coal. Mining machinery was expensive, and at first only the largest coal companies could afford the investment. By 1910, however, about half of the coal in the Big Sandy region was cut by machines rather than handpicks. Although these machines helped cut coal from the mine face, it was still necessary to blast and load the coal by hand in most mines, and miners continued to be paid by the carload.[16] Machines were also used to transport miners in and out of the mines. In the 1890s, electric locomotives began to replace the mules that hauled coal from the mines. The locomotives hauled low passenger cars called "man trips" into the mines.[17]

Timber setting machine used at Gilbert, West Virginia, 1944. Norfolk & Western Historical Photograph Collection, Virginia Polytechnic Institute and State Universities Libraries.

Technology also changed the way that areas were prepared for mining. Until the 1930s, timber beams were used to support the roof of a room. Later, roof bolts, sometimes called skyhooks, became more common. These bolts were long rods of steel driven into the rock above the roof to prevent its collapse. Other equipment that helped streamline the mining process included pumps that removed water from mine sites, telephones, rubber-tired buggies that replaced locomotive cars, conveyor belts that transported coal out of the mines, and massive fans for ventilation. By the 1920s, large mining sites had many structures, including steel tipples, powerhouses, fan houses, chutes, and coal bins. Bathhouses, lamp houses, and repair shops also became common. By the 1930s, a typical mine site in the Big Sandy region had electric cutting machines, electric locomotives for haulage, electric drills, lights, fans, and pumps.[18] For the miner, automatic loading machines were the next big change. Although introduced in the 1910s, they did not become common until the 1940s. Miners could now load many more tons of coal and were required to do much less hand loading. As a result, miners were no longer paid by the ton but began to receive wages for their labor. Automation greatly reduced the need for manpower, and the number of coal miners in the Big Sandy region dropped sharply after the 1940s.[19]

Postwar Mechanization

By the 1950s, far fewer coal miners were needed, productivity had increased substantially, and mining became a much safer occupation. The "continuous miner" had a great impact on coal mining. This machine clawed coal with large cutting teeth and pushed it away from the face. The coal was moved onto conveyors and taken out of the mine. Continuous miners replaced large numbers of workers, eliminated the need for explosives, and greatly increased productivity.

Management also changed. In the late nineteenth century, many mines were owner operated. Early owners often lived in the community, overseeing the entire mining process. They kept the books, paid employees, ran company stores, and knew their workers. As large corporations took over the industry, new management systems were introduced. Companies placed superintendents at a mine site to oversee the entire area and enforce company policies. Pit bosses, or mine foremen, supervised the day-to-day operations of the mines. They dealt directly with employees and were responsible for production and safety. At large mine sites, operations were further divided among section foremen, shift foremen, etc. These new management systems more closely supervised and regulated miners.

By the 1950s, new roads increased the number of small coal mines. Thousands of small truck mines, where coal was loaded into trucks and hauled away, appeared throughout the region. They were less mechanized and employed fewer miners. Many continued to use hand labor for the digging and loading of coal. They usually paid miners lower wages and were not well regulated. (Small mines continue to be an important source of coal in the Big Sandy region today.)[20]

Regulation

For most of the early twentieth century, the bituminous coal mining industry was poorly regulated. State governments were highly influenced by coal operators and large corporations. Heads of large coal companies, members of the boards of directors, and major stockholders worked closely with politicians, protecting the industry from regulation and unionization. West Virginia's Sen. Clarence Watson was the president of Consolidation Coal Company, the largest coal producer in the state. Many of Kentucky's coal operators joined the state's Democratic Party. Even John C.C. Mayo once considered running for office. Many of the editors and attorneys who served in state government had worked closely with coal companies.[21]

The only federal statute that governed mining in the nineteenth century was passed by Congress in 1891. It offered minimal recommendations for safety and ventilation and prohibited children under twelve years of age from working in mines. In the late nineteenth century, State Departments of Mines were established in Kentucky, West Virginia, and Virginia, but they seldom had the power to enforce laws. West Virginia passed laws defining the responsibilities of mine foremen and specified certain safety conditions in 1908. In Kentucky, Charles Norwood, the state's first mine inspector, led the campaign for mine safety and an 1884 law provided for the inspection of mines and offered minimal safety guidelines. But state agencies could only make recommendations and levy occasional fines. Mining was seen as a voluntary act: workers were taking part in a hazardous occupation and were regarded as responsible for most of the mishaps. Individual miners were often blamed when explosions, mechanical problems, or collapses occurred.[22] Coal operators fought safety legislation, fearing that individual states would lose their competitive edge. Some, such as West Virginia's A. Brooks Fleming of the Fairmont Coal Company, lobbied for federal mining safety laws that would affect all states. In 1910 a federal Bureau of Mines was created, but its main purpose was to research problems in the industry and make recommendations. The bureau issued guidelines for the coal mining industry, but it, too, did not have the power of enforcement.[23] The union battles of the 1910s brought new attention but few changes to the dangers of mining and living conditions in company towns. The U.S. Coal Commission's report in 1925 was quite critical of living and working conditions in many coal company towns, but it did not spark federal legislation. There were no further federal actions until the 1940s, when the Bureau of Mines was given the authority to enforce some basic safety standards. In 1952 the Federal Coal Mine Safety Act extended this authority by making some underground mines subject to annual inspection and providing penalties for coal operators who refused inspectors' requests or violated basic safety standards. In 1966 the authority to inspect was expanded to include all underground mines in the United States. States increased mine inspectors' authority as well in the 1950s, giving them the power to shut down mines that failed inspections.[24]

The Federal Coal Mine Health and Safety Act of 1969 strengthened many safety standards in the coal mining industry. It provided for the better ventilation of mines and called for reduced concentrations of coal dust, better roof supports, and mining equipment. The legislation was particularly important because it addressed the problems of black lung, or pneumoconiosis, a common health problem among miners, and provided compensation for miners afflicted with the disease. Individual states made further provisions. West Virginia and Kentucky expanded the powers of mine inspectors in the 1970s and established black lung funds. In 1977 the act was amended and expanded. Miners' rights were strengthened, safety standards were revised, and inspection was broadened. Coal mining is a much safer occupation today than it was earlier in the century. The federal Mine Safety and Health Administration now inspects mines and enforces mining safety standards, and numerous state and federal agencies oversee how coal is mined and regulate the environmental aspects of the industry.

While all these changes were taking place, more than 100,000 coal miners lost their lives from accidents and disease during the twentieth century.[25]

In recent years, new issues have emerged in the coal industry. Much of the debate about mining now focuses on its environmental impact and regulation of the industry. Regulation is often seen as a threat to areas where coal mining dominates the economy. As regulation has increased, miners and coal companies often find themselves on the same side, fighting for the protection of the coal industry in an effort to save jobs. New methods of mining, such as mountaintop removal and strip mining, have overshadowed debates about health and safety.

Coal Towns

For most of the nineteenth century, the Big Sandy River Valley was a remote, isolated part of Appalachia where a few small settlements allowed farmers to trade or sell what they produced. These settlements were usually located along a river or at a trail crossing where there was a store, mill, or market. When railroads entered the valley, new types of settlements developed. Temporary camps appeared for the men who were building the railroad or preparing sites for mining. Coal operators built entirely new towns in the wilderness to house their workers. Some of the small trading settlements grew into regional commercial centers serving the new industries and workers. Most of the towns listed in this guidebook were established during the coal and railroad boom around the turn of the twentieth century.[26]

Coal Camps

Often haphazardly thrown together and primitive, coal camps were squalid places. They were unplanned, temporary, mobile settlements filled with tents and shanties and crowded with young native mountaineers, African Americans, immigrants, and even convicts. Saloonkeepers, gamblers, and prostitutes followed the camps, and a violent frontierlike atmosphere often prevailed. Coal camps soon gained a reputation as dreadful, dangerous places offering little comfort after long hours of backbreaking labor. When the track was laid or the coal was depleted, the camps were abandoned. The shanties and dilapidated structures thrown up at the sites all disappeared long ago; today there are only ruined foundations or occasional debris in these locations.[27]

Company Towns

Once railroads opened the region, a great coal mining boom followed. Coal operators developed mine sites along the railroad tracks and for four decades, from the 1880s to the 1920s, hundreds of coal towns appeared throughout southern Appalachia. These coal towns were an improvement over the camps but still offered only the most primitive living and working conditions.

Many were built by local coal operators who had to economize. After investing money in rail lines and mining equipment, there was little left to spend on housing or conveniences. These coal towns were primitive, congested places that offered little more comfort than earlier camps. Monotonous rows of poorly constructed houses were built with timber cleared for the railroad. Hogs dug through the garbage, and waste and

The coal company town of Red Ash, Kentucky. Courtesy of Appalachian Photographic Archives, Alice Lloyd College.

runoff ran into the creeks. Chickens, ducks, and geese roamed the unpaved streets, and stagnant pools were filled with trash, mine waste, and debris. Well into the 1930s, the Big Sandy River Valley was filled with many poorly built, unhealthy, unsanitary coal towns.[28]

The narrow valleys and ravines of southern Appalachia were difficult places to build towns, and the natural landscape often dictated design and construction. Placed on the valley floor alongside the rivers and creeks, these towns usually had a linear design following the winding, meandering path the waterways had carved out over the years. Railroad tracks, houses, and mine equipment were laid out. Stores, depots, machine shops, and mine sites were strung out along the railroad tracks. When all of the available land was used, houses were built on the surrounding hillsides.[29] Coal towns were crowded with distinct residential, commercial, and industrial areas. The congestion, dirt, and noise prevented these towns from developing the romantic image that farming communities had in the late nineteenth century.[30]

The arrival of large corporations from the North led to some improvements. Standardized housing was constructed by local laborers. Contractors who specialized in building houses for mining companies were hired to build new towns. These new company coal towns, usually located in remote, isolated locations, offered basic housing for miners and their families but still lacked many conveniences. Most dwellings were simple detached, wood-frame houses in a square or rectangular design, with three or four rooms, a porch, and an outside privy. There was a large increase in the number of company coal towns during the early twentieth century. By the 1920s about 70 percent of coal miners in the country lived in company towns. In the Big Sandy River Valley the percentage was even higher, approaching 80 percent.[31] Reporters, union organizers, and social workers from the North described them as desolate and dirty places. They were cited as having some of the worst living conditions in the nation, along with continued problems of violence, prostitution, and poor sanitation.[32]

In most coal towns, miners and their families lived a hard life in substandard housing. Coal operators invested little in these settlements, and when the coal was depleted, the towns were abandoned. Equipment was moved, the houses were dismantled or used for firewood, and the towns disappeared. Most of the coal towns of the Big Sandy region have long since deteriorated and no longer exist. Today, when driving through the

Big Sandy River Valley, there are often small clusters of older homes that are similar in appearance spread out along a creek or railroad track. In many cases, these are all that remain of a once bustling coal town.[33]

Although coal towns were poor places to live and work, they drew countless thousands of people. Mining was the only work available, and it offered one of the few ways to earn wages in these rugged mountainous areas. The populations of remote rural counties soared and new towns appeared throughout the hills. Even though housing in a typical company town was primitive, it was usually better than what was available in many of the surrounding Appalachian communities. Whitesburg, Kentucky, attorney and author Harry Caudill explained: "To the mountaineers in whose midst such houses sprang up they were palatial, and rumors spread with wildfire speed throughout the plateau that 'fine houses' were being built by the hundreds. The mountaineer had never experienced such quality construction and few of them had ever so much as seen a plastered wall. Compared to his cabins and crudely built frame houses, the residences constructed by the large companies were indeed enticing."[34] Houses that today appear to be shabby and poorly constructed were often a step up from the simple log homes of the region. Even if housing was not particularly attractive, coal mining towns offered services such as company stores, churches, clinics, and recreation halls that drew miners and their families.

Model Company Towns

In the late nineteenth century, a few large coal companies built model company towns, towns that offered quality housing, sanitation, and recreational facilities. By offering an improved standard of living, planners hoped to attract stable workers with families, thereby controlling many of the problems that plagued early camps. Model company towns were an effort to create an ideal industrial community where carefully planned streets with high-quality housing and conveniences would offer a suburban lifestyle to people living in the most remote mining areas. Only the largest corporations could afford to build such communities, but they hoped that by providing a good living for their workers they would improve the image of the coal town. The earliest model company coal towns were Virginia Coal and Iron's Stonega, Virginia; U.S. Oil's Holden, West Virginia; and U.S. Steel's Gary, West Virginia. As more

The main street of Jenkins, Kentucky, one of Consolidation Coal's model company towns. Jenkins Collection, Mary Jo Wolfe Memorial Library.

model company communities were built after 1910, competition developed and even small coal operators invested more in their towns. The profits of the First World War allowed even greater expansion.[35]

Model company towns stood in stark contrast to the typical settlements of the region. They were a small portion (less than 2 percent) of the coal towns of southern Appalachia but soon developed great reputations. Model company towns drew people from the surrounding communities with their beautifully landscaped streets, parks, and recreation facilities and modern conveniences. Holden, West Virginia, had a modern movie theater, library, clubhouse, bowling alley, and squash court. Jenkins, Kentucky, had a sewer system, company dairy and butcher, large company stores, and regular garbage collection. Appearance was a high priority, and homes in model company towns were more attractive, well maintained, and frequently upgraded. These towns quickly filled to capacity and often had long waiting lists of employees who wanted to move there with their families.[36]

The leading industry journal, *Coal Age*, focused on mining, technology, and preparation but also served as the public relations organ of the coal mining industry, spreading good news about the lifestyle that big

Downtown Wheelwright, Kentucky, Inland Steel's model coal town in Floyd County. Russell Lee Collection, National Archives Still Pictures Branch.

coal companies offered their employees. The journal regularly celebrated the life of a miner in a model company town. Managers from large corporations such as Consolidation Coal wrote articles promoting the company's commitment to modern housing, healthcare, community services, and benefits. Model company towns were not common, but they were often of notable size. Communities such as Jenkins or Benham, Kentucky, were typical with thousands of residents.[37]

The model company town was partly a reaction to the threat of organized labor. In many of the labor battles of the early twentieth century, union leaders cited poor housing and a lack of sanitation and services in their attacks on coal companies. Coal companies wanted to show that they took good care of their workers and could ultimately offer them better benefits, wages, and protection than a labor union. This system of welfare capitalism protected workers and allowed managers an opportunity to mold personal behavior and teach proper values. They hoped that this would make employees fiercely loyal to the company and discourage union activities.[38] The promise of a good life in a model company town also allowed coal companies to recruit a diverse workforce of white mountaineers, African Americans, and European immigrants. The Big Sandy

Valley soon had large populations of Germans, Hungarians, Italians, and Poles. By the 1910s almost every European ethnic group was found in the Big Sandy region. Many miners living and working in nearby coal towns moved to the new model company towns with their families.

Visitors to a model company town will notice several differences. They were designed and laid out by professional planners who emphasized space, landscaping, and uniformity similar to that of the new suburban developments of the north. Many featured parks, sports facilities, landscaped streets, and quality schools. The recreational, medical, and educational facilities are usually the most prominent remaining structures today. In model company towns, houses received special attention and were often painted, repaired, upgraded, and remodeled. Many of these houses are still occupied today, decades after the coal companies abandoned them. Some of the best preserved model company towns are Benham and Lynch, Kentucky. Jenkins and Van Lear, Kentucky, and Gary, West Virginia, have far fewer surviving structures but were once communities of the same size and design. Stonega, Virginia, and some smaller surrounding towns such as Dunbar and Derby were also promoted as model company towns in the early twentieth century but have recently suffered great structural losses. In the 1920s the development of company towns slowed and the Great Depression brought the era of the model company town to an end.[39] One of the last model towns was Wheelwright, developed in the 1930s.[40] When visiting model coal company towns in the Big Sandy region, despite their state of decay or loss, travelers will notice that houses are farther apart, streets are wider, and buildings are often more architecturally attractive.[41]

Regional Commercial Centers

Regional commercial centers existed before the coal boom but grew and prospered by serving nearby mining communities. In the mid-nineteenth century, when southern Appalachia was a land of small, self-sufficient farmers, such as the Hatfields and the McCoys, regional commercial centers served as marketplaces. They were located along the rivers, allowing them contact with other communities in the Ohio Valley. A system of roads and turnpikes connected the Big Sandy region with the outside world, but transportation was greatly limited. Many regional commercial centers were county seats. Other regional commercial centers developed near mine sites, offering services not available in company towns. Some developed bad reputations as places where drinking, prostitution, gambling, and violence were common. The relationship between regional commercial centers and remote mining communities was complicated. As centers of county government, some were responsible for maintaining police forces and school systems and enforcing laws. Even though coal towns produced great profits, they seldom contributed much to the local tax base. Coal companies often built their own schools and had their own police and security forces, giving them a great deal of independence. When conflicts erupted between coal operators and union organizers, local authorities were caught in the middle but ended up siding with the coal companies. The relationship between coal towns and regional commercial centers is an area of scholarship that deserves much more attention than it has received. The centers played a critical role in the development of the region and were the site of many important events.

Regional commercial centers appear much different than coal towns. They have often undergone several periods of development and are spread along the banks of rivers and streams or along railroad tracks. Today they are located on major highways. They have distinct downtown commercial districts with stores, banks, and government agencies. The county courthouse is usually the most prominent building in the town. Once the site of coal company offices and the home of many coal operators, the regional commercial centers have a wide variety of houses from different periods. Although the coal booms have come and gone, these towns continue to be centers of commercial, political, and social activity. Pikeville and Paintsville, Kentucky, and Williamson and Welch, West Virginia, are good examples of regional commercial centers in the Big Sandy region.

Sites and Structures Found in Coal Towns

There are many different types of buildings, remains, and ruins in the Big Sandy River Valley. It is often difficult to tell that a settlement was once a booming mining town because the structures directly related to mining and railroading have been lost over the years. As coal towns were abandoned, coal companies sealed mines, dismantled tipples and coke ovens, removed machinery, and destroyed nearby buildings. Railroad tracks were dismantled, abandoned, or overgrown. In many cases, rows of company housing were demolished and mining equipment was disassembled and moved. Only a few prominent buildings and structures have been designated as historic landmarks or placed on the National Register of Historic Places. Even so, a careful observer can usually find some small traces of the mining past in any coal town.

Houses

The most common surviving buildings in Big Sandy River coal towns are houses. Those built after 1890 are much more numerous, and many examples can be seen throughout the region.[42] When most coal companies built a town, they needed simple, inexpensive houses where workers and their families could live. They used cheap, low-quality materials and built basic functional dwellings. Houses in southern Appalachian mining towns gained a reputation for their shabby, poorly built homes and unsanitary conditions. They were often painted the same drab color, if they were painted at all. They were typically one-story boxlike houses placed on post foundations. A pot-bellied stove placed in the center of the home provided heat. Electricity, usually a single bare bulb in the middle of a room, became more common by the 1910s. They were crowded into available land, usually right up against the railroad tracks and the creeks. Houses nearest the tipple were coated with coal dust. Some coal town houses were so bad that people mistook them for chicken coops or stables. Well into the twentieth century these towns were still referred to as coal camps.[43]

More than 90 percent of coal company houses were box frame, where a simple square lumber frame was built, usually of pine or hemlock. Interior walls were bare wood. Running water or indoor toilets were rare.[44] Electricity was usually the first improvement made, brought to many towns in the 1920s. Many of these houses were so poorly built that in 1916 the Bureau of Mines issued a series of guidelines for the construction of houses in mining towns. When the U.S. Coal Commission surveyed America's coal company towns in 1923, it still found large numbers of dwellings that were considered substandard.[45]

Miners' houses were generally small with few amenities. In 1917, at the peak of the coal boom, *Coal Age* cited two common housing designs. The standard three-room, one-story house measured twenty-four by twenty-four feet. It had a rectangular front room off the porch and two

Miners' houses under construction in Burdine, Kentucky. Jenkins Collection, Mary Jo Wolfe Memorial Library.

twelve-by-fourteen-feet rooms in the rear. The front room served as a living room, and in back were a kitchen and dining area and a bedroom. A second, slightly larger, four-room design (twenty-eight by twenty-eight feet) had a second bedroom. Where space was limited, shotgun houses were also built. These were low, small, freestanding clapboard homes placed closely together on long rectangular lots. Two or three rooms, the full width of the house, opened directly on to the next. There was a gable-end entry and a front and back porch. They were called shotgun houses because if a gun were fired through the front door, the shot would pass through all of the rooms and out the back door. There was sometimes a second floor added at the rear of the house.

Two-story houses were also common. In narrow hollows where space was limited, a popular design was known as the Basic I. With this design, the house had an end-facing gabled roof and was two rooms long but only one room deep, allowing it to fit on a narrow lot up against a mountainside. A Basic L design was also popular, along with a two-story shotgun style.

Two-story shotgun houses in Seco, Kentucky.

There were some changes by the 1910s. Early coal towns were built entirely of local materials, but later large companies, such as Consolidation Coal, contracted builders to construct low-cost housing that could be assembled on site with few additional materials. In the early twentieth century some of the largest contractors were Bay City Ready-Cut Homes of Michigan and the Nicola Building Company of Pittsburgh. They sold prefabricated homes or home kits ranging from $500 to $1,500. Consolidation Coal purchased many three- and four-room homes from Nicola for Jenkins and Van Lear, Kentucky, for about $600 each.[46] As prefabricated housing became more common, another style known as the two-story four pen appeared. This was a semidetached two-family home with two entrances, front and back porches, and five rooms for each family. Downstairs there was a living room, kitchen, and dining area. Upstairs there were two bedrooms. The two-story four pen was most common in the Flat Top–Pocahontas coalfields of West Virginia and Virginia but is found throughout the region.

A few coal companies realized that well-maintained homes were an investment, especially at sites where the mines might be productive for many years. Good housing could produce income and yield a profit over the long term. Coal companies usually leased land in coalfields and therefore could not actually sell property to their employees, so most miners did not own their homes. Instead, they were charged rent, which was deducted directly from their pay. In the 1920s, the average rent was between $1.50 and $2.00 per room, per month, along with deductions for electricity and fuel. With mining attracting many single men, enterprising families would rent larger homes and take in boarders as a way to supplement family income. The house's primary tenants paid rent to the company through payroll deductions, and boarders paid the family directly.[47]

Even though the largest coal companies tried to improve housing for their employees, most did not. A survey of coal company houses in the early 1920s found less than 15 percent with running water and only 1 percent with indoor baths, showers, or toilets. In 1949 a similar study reported some improvement, but still only half of coal company houses had running water and only one-quarter had indoor plumbing.[48]

Houses in Model Company Towns

Houses were of considerably better quality in model company towns. Large coal companies such as Consolidation Coal or Island Creek Coal made showplaces of their mining town homes and offered amenities and services comparable to cities in the north. In *Coal Age,* companies offered

Well-kept miners' homes in Benham, Kentucky, circa 1920. Reprinted from *Coal Age* (1921).

detailed descriptions of these houses and the materials used to construct them. They boasted of well-manufactured hemlock and pine stock that was used in the framing. Gypsum, wood fiber, and plaster walls lined the interiors. Hard brick fireplaces and chimneys, composition roofing, and spacious front porches were standard in their homes. They were placed on larger lots and had areas to grow gardens. Miners and their families enjoyed many conveniences, including electricity, running water, coal delivery, and garbage collection. Even indoor toilets were added by the 1920s. Houses in model company towns were built to higher standards and are more likely to still be in use today. Homes in Holden, Jenkins, Benham, and Lynch are typical, and many have been well maintained and renovated over the years.

Boardinghouses

Many companies also built boardinghouses for single miners or married men who had not yet brought their families to the community. Boardinghouses are not easy to identify, since they vary in size and construction quality and are typically made with the same materials used for houses. Because there were usually only a few in a coal town, boardinghouses lack the uniform architectural style of company houses. Sometimes large duplexes or fourplexes were modified for boarders. In towns with African American communities, separate boardinghouses were built for white and black workers. When visiting coal towns, it is often difficult to distinguish larger homes from boardinghouses.

Coal Operators' and Managers' Houses

Coal operators, supervisors, managers, and company officials lived in entirely different houses than the miners. Their homes were larger, better built, less standardized, and less uniform in appearance. Early coal operators, who often lived in their coal towns, built spacious residences for their families. These were the largest, most elaborate dwellings in the town. Some were three stories tall with large expensive windows, steep gabled roofs, and deep wraparound porches. They were placed on spacious, well-kept grounds, often on a hillside overlooking the town. These houses were built with higher quality materials. Cut stone, wood shingles, slate roofs, and brick were commonly used. The interiors were also elaborate.

Houses built for managers and company officials in the lake district of Jenkins, Kentucky.

High ceilings, winding staircases, beautifully carved woodwork were seen throughout. They were outfitted with all of the latest conveniences: electricity, indoor plumbing, and a range of appliances. There were large parlors and dining areas for entertaining. However, as large coal companies took over the towns, coal operators were replaced by superintendents and managers. Their homes were still large and elaborate but much less elegant.

An assortment of styles and designs define these residences, which were built in various lots around town. Most were bungalows with full porches and pitched roofs and gables; many were late Victorian style; and some were custom made, featuring eclectic elements from many styles. In some cases they were simply larger and more elaborate versions of company housing. In Jenkins, an entire lakeside district of beautiful homes was built for supervisors, managers, and company officials. In many coal mining towns these houses reflected the superior position of the officials and were placed on hillsides, above miners' homes, or in a very prominent place in the center of town. Sometimes visitors climbed the stairs and were required to enter through the back door. Coal operators and managers often wore formal dress at home, further emphasizing their positions of authority.[49]

Segregated Housing

Throughout the Big Sandy River Valley company housing was segregated. Although much coal town housing was low quality, homes for African American workers and their families were usually inferior to those of whites. Black miners were placed in separate areas of town, in poor-quality, old, damaged structures. Cheaper materials were used for these homes, and they were not finished in the same way or regularly upgraded. When repairs were needed, African Americans were placed at the bottom of the list. There were instances where all of the homes in an entire town were improved, except for those of black miners. African American miners' houses were also crowed with several families and occasional boarders. Even if housing was of the same quality, black workers lived in the least desirable locations. They were placed near tipples, railroad tracks, or coke ovens, where there was considerably more noise and pollutants.[50] Even so, for many African Americans, conditions improved as they moved into coal towns. Better-quality housing was being built throughout the region, and by the 1920s some African American miners were able to leave company housing and purchase homes in nearby towns. Neighborhoods were still segregated, but home ownership gave blacks an opportunity to purchase better houses and develop independent communities. Williamson, Welch, and Keystone all had notable black neighborhoods by the 1920s. This process slowed in the 1930s as the Great Depression caused many miners, both black and white, to lose their homes.[51]

Churches

The young single men who filled the early coal camps showed little interest in religion. Itinerant preachers sometimes visited the camps and built makeshift structures where they could preach. Churches became more common as coal companies recruited married workers and their families. In coal towns, they were built and supported by the mining companies to serve the spiritual needs of the community and to be a place for families to socialize. But they were also an effort to bolster the moral image of coal companies and influence the behavior of town residents. Managers hoped that activities at churches would help curb drinking, gambling, and violence. Ministers were expected to promote the company and could be fired if they took an interest in wage disputes, personnel matters, or union organizing activities. Large coal corporations included churches in their original town plans and encouraged membership. Consolidation Coal built large, prominent churches at Jenkins and Van Lear. Stonega Coke and Coal provided churches for its towns in southwest Virginia, usually adding one for each new colliery. U.S. Steel built many churches in Lynch, Kentucky, and Gary, West Virginia.[52]

Churches are some of the most common buildings remaining in coal towns today and many are still in use. They were usually built from the same materials as houses and were simple wooden frame buildings with front and rear facing gables, a cupola for a bell, or an attached bell tower. They sometimes had cathedral type windows and a chimney, usually at the rear. Many were built as nondenominational churches to serve the needs of various religions. Others were built as Methodist or Baptist churches, the major religions in the Big Sandy region. Some of these churches were large brick structures that dominated the center of town with elaborate stained glass windows. Separate, smaller, simpler churches were built for African Americans. Because of large immigrant populations

Catholic Church in Jenkins, Kentucky.

in the early twentieth century, Catholic churches were common in many coal towns. Occasionally, Russian or Greek Orthodox churches were built in communities that had large eastern European populations. Many of these churches have also survived and continue to have active memberships. Van Lear and Jenkins, Kentucky, have large Catholic churches. Elkhorn, West Virginia, has a beautiful gold-domed Russian Orthodox church that remained in use until the late 1990s.[53]

After the collapse of the coal industry in the 1920s and the Great Depression that followed, coal companies began to limit services in mining towns. In the 1940s and 1950s they sold most of their properties, and the churches were taken over by the community. Church buildings can be a clue to the former existence of a coal town even though other buildings have long since disappeared, and they can provide clues to the ethnic layout of mining communities. Protestant churches served the southern white population and management and were usually located in the center of town. Catholic churches were built in immigrant neighborhoods, and Baptist churches were often built in the African American section of town.

Schools

By the early twentieth century, most states had compulsory education laws, and coal companies built schools to provide miners' children with a basic education. In West Virginia, children between the ages of seven and fourteen were required to go to school. Those who were fourteen or fifteen and not regularly employed were also expected to be in school. Coal town schools varied tremendously in size, quality, and design. Large model company towns built beautiful schools and recruited qualified teachers. These buildings were some of the most prominent structures in town and served as meeting places and social centers. One of Benham, Kentucky's, stone and brick school buildings has recently been restored and opened as an inn. In Jenkins, Kentucky, the original school, a two-story brick building on the town's main street, continues to house offices. In more typical company towns, schools were basic and functional; sometimes classes were simply held in an unoccupied home or vacant building. In small communities, a one-room schoolhouse that all students attended was more common. These were simple wooden frame structures with a coal stove for warmth and benches and tables for instruction. (Some of the coal towns of the Big Sandy River Valley had

The ruins of a school building constructed as a WPA project in Shelbiana, Kentucky.

one-room schoolhouses that were used until the 1970s.) In most coal towns the county provided teachers, who were housed in small dwellings known as teacherages, and sometimes the coal companies supplemented their meager salaries.[54]

Under the New Deal programs of the 1930s, hundreds of schools were constructed or remodeled as Public Works Administration (PWA) or Works Progress Administration (WPA) projects throughout southern Appalachia. Many of these projects employed skilled stonemasons and produced beautiful buildings made from quarried stone with intricate detail work. Native sandstone and limestone were the most common materials used. Others were made of brick, built to replace dilapidated wooden structures or designed to consolidate several local schools into one large facility. Many of these buildings are still in use today as schools, offices, or support service centers. Schools built under the WPA continued to segregate black children, constructing large new schools for the white population and smaller, basic facilities for African Americans. Schools built for African Americans can sometimes be identified by lettering or plaques designating them as "colored" schools, and some look like large houses. Smaller school buildings in the area have been renovated and today serve as private residences.[55]

Company Stores

The center of activity in a coal town was usually the company store, also known as the commissary. It was located in the center of town, along the railroad tracks, and carried necessary consumer goods and sometimes exotic luxury goods from faraway places. Many miners had to purchase their own work supplies, which were readily available at company stores.[56] In small company towns, stores carried only the bare necessities: clothing, mining supplies, food, and drugs. In large towns, company stores resembled contemporary department stores, featuring different sections for men, women, and children, home furnishings, and entertainment. Some of the more elaborate establishments had soda fountains, barbershops, beauty parlors, and shoe shine service. By the 1920s company stores expanded to service automobiles, offering gasoline, auto parts, and basic repairs.

Much has been written about companies stores because of the scrip system. Coal companies paid their employees with scrip, paper money or coins printed by the company, in place of cash. Miners were paid at the company store, usually twice a month, with deductions for rent, utilities, medical fees, and advances. Company bills or coins could be used at company stores or at neighboring businesses. The scrip system was based on dollars, but scrip was usually discounted about 25 percent when converted to currency. Many companies paid in both scrip and cash. There is a popular impression that miners were bound by debt to company stores. They paid inflated prices, had few other places to shop, and took frequent advances on their pay. In "Sixteen Tons," a popular song of the 1950s, a miner sings, "I owe my soul to the company store." This was probably the case in the isolated coal camps of the nineteenth century, but by the 1910s there was regular rail service to most coal towns, and miners could frequent merchants in nearby towns. Scrip was common, but many miners had cash and were seldom in debt. The automobile made this even easier, and in thriving regional commercial centers such as Pikeville, Kentucky, or Matewan and Williamson, West Virginia, goods were available at competitive prices. Company stores did charge slightly higher prices but were usually used for convenience.[57]

Company stores were also a place to socialize. It was years before most company houses had refrigerators, so women made daily trips to shop for food. While orders were being filled or bills prepared, they had an opportunity to visit with neighbors. Men often sat outside the company store

Wisconsin Steel's company store building now houses the Kentucky Coal Mining Museum in Benham.

greeting patrons and discussing the day's events. On weekends in some of the larger company stores, a family could spend the better part of the day socializing, shopping, or being entertained. In towns with significant African American populations, company stores were segregated. Some had separate sections for black customers, while in other towns there were completely separate stores. European immigrants were usually welcome at white facilities but were not necessarily treated equally.[58]

The Company store was one of the most prominent buildings in coal towns. It often dominated the center of town, some facing the main street with the backside along the railroad tracks. It was often a three-story rectangular building made primarily of wood or red brick with flat or low-angled roofs. These tall buildings were usually placed on a solid foundation of stone, brick, or concrete. The first level was used for shipping, receiving, and storage. Steps led up to a second level where retail space was located. The third level was usually for storage or offices. Large display windows showcased new merchandise.[59] In model company towns, the commissary was a source of company pride and a symbol of its commitment to its employees. Companies built large elaborate stores designed to compete with merchants in nearby communities, and they boasted that their stores offered all of the conveniences found in nearby cities at competitive prices. Most coal company stores are gone, but some remain in use today. In Van Lear, Kentucky, a former Consolidation Coal commissary is one of the town's local convenience stores. In Benham, Kentucky, International Harvester's spacious brick and stone company store has been turned into a coal mining museum with displays that include merchandise, tools, and scrip.

Mines

Thousands of mines have operated in the Big Sandy region and most were abandoned long ago. Since the 1960s, much of the coal in the Big Sandy region has been strip-mined. In drift and shaft mines, the exterior landscape remains basically intact and, to the casual observer, it is sometimes difficult to determine the location of these mine sites. In strip-mining, the surface of mountaintops is literally sheared away and seams of coal are exposed. Strip-mining is usually carried out where coal veins are close to the surface and earth-moving equipment such as bulldozers and power shovels can be used to remove the overburden. Until 1977 sites that were stripped were often left bare, creating serious problems with run-off waters and erosion. The 1977 Surface Mining Control and Reclamation Act now requires that lands be returned to the original contours as closely as possible and that vegetation be replaced. Strip-mined sites are not always easily seen by visitors to the Big Sandy region. Many are located on mountaintops, quite visible from the air but difficult to see from below. In some areas, such as Peach Orchard, Kentucky, or Pardee, Virginia, entire towns have disappeared due to strip-mining. The entire town of Blair, West Virginia, an important site in the 1921 Battle of Blair Mountain, is presently threatened by strip-mining.

Today, abandoned mine sites can be found along railroad tracks or former railroad grades. Major mines such as those at Lynch or Wheelwright, Kentucky, are easy to locate and frequently have signs or markers identifying them. Some have evidence of the tracks that hauled men and coal in and out of the mines. Some mines have been blasted to collapse the entrance or are cemented or barred shut. Although most major mines have been sealed, many have simply been abandoned. Sometimes there are coal cars, rail ties, and loading equipment abandoned near mine sites. Entering an abandoned mine can be extremely dangerous and also constitutes trespassing. However, several towns, such as Lynch, Kentucky, Pocahontas, Virginia, and Beckley, West Virginia, have mines opened to the public for safe viewing.

Tipples

Early coal tipples (known as breakers in northern coal mining districts) were simple wooden structures located alongside, or directly over, railroad tracks. After coal was dumped into the tipple it was channeled through a chute into railroad cars below, also known as hoppers. In early mines, workers hand sorted coal in the tipple. They separated large and small pieces and removed slate, rock, and debris from "dirty coal." As tipples became more sophisticated, a series of screens sorted coal by size. It would pass through the holes and separate into different storage areas. The sizes determined the grade of the coal and its use. Coal stoves and home furnaces used very small pea-sized coal. Industries used larger lump-sized coal. After sizing and grading, workers sometimes washed the coal. Washing removed fine dust and debris. Oily sprays kept the remaining coal dust minimal.

Near the tipple were mountains of discarded slate and low-grade coal. Some of these piles were hundreds of feet high and were susceptible to spontaneous combustion. These "slate dumps" could burn for days and created oily black smoke and nauseous fumes. This added to the already serious pollution problems in coal towns of locomotive smoke, burning coal stoves, and coal dust.[60]

Once the coal was sorted, graded, and washed it was ready to be loaded. A railroad hopper was placed under the tipple, the chute was positioned (sometimes by hand), and the coal was released into the car. Workers spread out the coal in the hopper and leveled the piles as more was added. Fully loaded coal hoppers were moved to side tracks to be taken away. The process of taking loaded coal cars to the railroad yards and returning empty hoppers was known as a mine run. During peak production there were many mine runs a day and often a shortage of empty coal cars.

As coal mining became more sophisticated, so did the tipple. Coal was automatically transported to the tipple and large shaker screens sorted the grades. By the 1910s coal tipples included washing sites and

The ruins of a wooden coal tipple at Ajax, Kentucky.

An unidentified coal tipple and preparation complex in the Big Sandy River Valley, circa 1940. Works Progress Administration Collection, Public Records Division, Kentucky Department for Libraries and Archives.

preparation plants. They stretched out over many railroad tracks and could load several railroad cars at once. The term tipple later came to include the entire surface structure of a mine, the preparation plant, and the loading tracks. Concrete and steel structures began to replace wooden tipples by the 1920s. The Lynch tipple, erected in the early 1920s and still standing today, was the largest built up to that time and was made entirely of concrete and steel.[61] As the process of mining became more mechanized, the tipple area became more complex. The tipple area included head houses, small offices, and preparation and railroad facilities. Advanced technology and the automation of coal loading made tipple areas busy industrial centers. Wheelwright's last coal tipple was state-of-the-art, using 1940s advances in electronic technology. Most of the older tipples in the Big Sandy River Valley were destroyed years ago, but the remains of wooden tipples can sometimes be found rotting alongside the railroad tracks. The only surviving wooden tipple of significant size in the region is the Ajax tipple in Perry County. Other tipples from the early decades of the twentieth century are still in use. Unfortunately, when mines are closed, the mine sites are usually the first to be dismantled or demolished.[62]

Coke Ovens

Many coal mining districts had rows of coke ovens where bituminous coal was turned into coke, a light and porous by-product of coal. The coal was baked at extremely high temperatures in firebrick ovens. Tar, gas, and other by-products were lost during the baking process, which made coke a high-grade fuel that could be used in smelters and iron furnaces. Until the 1930s, most coke was produced near the mines in beehive ovens, named for their spherical shape. Beehive ovens are open at the top and have small openings for draft along the base. Made of stone or brick, the ovens were arranged in banks or rows along railroad tracks. Many laborers were needed to run a bank of coke ovens, which could contain up to four hundred ovens. In the 1910s, Stonega, Virginia, had a coking crew of more than 250 men. Because coal was loaded into the ovens and coke was pulled or unloaded by hand, temperatures near the ovens were quite high. Loading and unloading coal and coke required as much strength as loading coal in the mines. Workers used large pitchforks to load coal into the ovens and remove it for placement into wheelbarrows. Then they

Coke ovens under construction at Keystone, West Virginia, circa 1900. Norfolk & Western Historical Photograph Collection, Virginia Polytechnic Institute and State Universities Libraries.

rolled the wheelbarrows to the tracks and dumped the coke into the hoppers. This was very demanding work and one of the least desirable jobs, often delegated to African American workers.[63]

In the 1920s, coking operations were moved to the cities and factories where new by-product ovens replaced small beehive ovens. Although there were thousands of beehive coke ovens in operation at the turn of the century, few remain today.[64] Coke ovens covered large areas and were often the first structures to be demolished. There are some stone remains at Stonega and a row of modern brick coke ovens at Pine Branch, Virginia. The only place to view modern by-product coking operations in the Big Sandy region is in Grundy, Virginia, where a series of ovens operate near the junction of VA 638 and U.S. 460, south of the city.

Many other mining structures can be found in coal towns throughout the Big Sandy region. Near the entrance to a mine there was often a head house, a small office where supervisors oversaw operations and equipment was distributed. Some larger mines had separate lamp houses where miners received fully charged headlamps and assorted equipment prior to entering the mines. Many larger mining operations had bathhouses where miners could clean up after work and change into street clothes. Bathhouses can usually be distinguished by large interior areas that were used for miners to change their clothes before and after work. Large concrete floor areas with drains often indicate where showers were located. Larger mining districts often had their own powerhouses that produced electricity from coal to run the mines and serve the town. Typically, powerhouses were brick structures, along railroad lines, with large smokestacks. An excellent example of a powerhouse is at Lynch, Kentucky. By the 1950s these powerhouses were abandoned and coal operators purchased electricity from utility companies.

Community, Recreation, and Leisure Facilities

There were various types of community and recreation facilities in company coal towns. In small settlements, a basic meeting area, sometimes part of the church or school, was used for community functions. Many coal towns had buildings for recreation that featured amusements such as billiards, films, meeting rooms for clubs or lodges, soda fountains, or even libraries. YMCA facilities were common in many larger coal towns in the early twentieth century. Community and recreation facilities were large, complex places in model company towns. They were designed to serve miners, local residents, and visitors. Jenkins, Kentucky, had a clubhouse for miners as well as a theater, soda fountain, and YMCA. Wheelwright, Kentucky, had a community center with a restaurant, library, movie theater, and bowling alley. Some offered lodging for visitors and entertainers. Lynch, Kentucky, had a massive hotel that featured fine dining, music, live theater, and prominent popular singers in the 1920s. Baseball fields and small playgrounds were also common, and many coal companies sponsored sports teams. Near community buildings there were sometimes swimming areas, tennis courts, and playgrounds. Community buildings and recreation centers were multipurpose facilities and varied greatly in size and design. They were usually large, over one hundred feet in length at times, and had a front porch or sheltered area. Seldom were the buildings attractive or architecturally significant. Inside, they often had duplicate facilities for segregated communities.[65]

Medical Facilities

Company towns usually provided basic medical services through a town doctor and a corps of nurses, deducting one to two dollars a month from workers' payrolls for health care. Large communities had clinics and hospitals that served the region and provided education as well as health care. Because they were placed in company offices or in buildings resembling homes, coal town clinics and hospitals are sometimes hard to identify. As companies cut services to employees after the Second World War, many company medical benefits were reduced or eliminated. In the 1950s, the UMWA opened a series of regional hospitals to serve coal miners and their families as the company facilities were closed. These large, modern structures were made of glass, concrete, and brick and were strategically placed throughout the region. Most are still in operation today as part of the Appalachian Regional Healthcare system.

Coal Houses

Coal companies sold coal to their tenants for use in heating grates and coal stoves. Small sturdy wooden or brick structures known as coal houses were built in the front or back of houses for delivery and storage of coal. Designed to protect coal from precipitation, coal houses had openings at the top for loading the coal and openings at the bottom for removal. They were typically six feet long, four feet high, and four feet wide and held about one ton of coal. Homes with two families had double-sized coal houses. Miners paid between $1.50 and $2.00 a ton for coal for home use in the 1920s. The most prominent coal houses still in existence today are those at Derby, Virginia, where red brick and block coal houses were designed to blend architecturally with company housing. Pocahontas, Virginia, also has some remaining examples.[66]

Coal houses in front of homes at Derby, Virginia.

Privies

Most mining towns in the Big Sandy region did not have indoor bathroom facilities until the mid-twentieth century. Wooden privies (or outhouses) were used in almost all of the coal towns in this guidebook. There were two basic designs: a single privy, about three to four feet wide and long and seven feet high, or double privies, which were about seven feet wide. A double privy had one or two stools, with an inside partition separating them. There was usually one privy for each family and others were placed throughout the town for community use. Privies were located along streams so that waste could be carried away from the town. They were not well screened, flies were often a problem, and the odors could be quite offensive at times. Few coal companies spent money maintaining privies, but they usually provided cleaning twice a year.

Indoor plumbing came to the Big Sandy region much later than other parts of the country. In the 1930s, when documentary photographers working for New Deal programs surveyed coal towns, they were shocked by the lack of bathroom facilities and often photographed homes and privies along the rivers and creeks. Today, privies from the early twentieth century are rare and indoor toilets are common everywhere. Once indoor plumbing was installed, families quickly, and proudly, destroyed their privies. Occasionally one can find ruins of privies along streams.[67]

Garages

As roads were constructed in the 1920s and 1930s, automobiles appeared in coal towns (often serving as taxis) and connected communities that the railroad did not reach. In much of the Big Sandy region, local roads were of poor quality, and personal automobiles were rare until the 1940s. After World War II, cars became more common and miners began to commute to work. Some companies constructed row garages near houses for miners to store their vehicles. They were usually made of wood and provided the minimum space needed to park an automobile.

Bridges

There are a variety of historic bridges in the Big Sandy region. The terrain among the hills and hollows made it difficult for people and vehicles to move freely from one location to another, and the building of mining towns along rivers and creeks led to problems with flooding and raging waters during heavy rains. Some of these bridges were simple structures built by local residents; others are quite elaborate and were built as public works projects. Many southern Appalachian communities have footbridges that span creeks and streams where rising waters are a problem. The main towers, beams, and planks are made entirely of wood, and steel cables stabilize and support the bridge. A network of wires or ropes act as handrails for people crossing the span. At one time, these bridges connected residential areas with mine sites or led from workers' homes to the main roads or rail lines. Today most of these bridges have been replaced or abandoned, but their remains can frequently be seen along the banks of rivers and streams, overgrown by brush. In a few locations, newer steel versions of these traditional bridges have been built. At River, Kentucky, a modern cable suspension footbridge spans the Tug Fork.

Some suspension bridges were built for automobile and truck traffic. On U.S. 23 north of Pikeville, the Boldman Bridge is a good surviving example of a simple swinging suspension bridge designed for vehicular traffic, common in the early decades of the twentieth century. A series of cables are connected to the wooden planks that form the bridge's base. Two towers constructed of I-beams and channels with massive concrete anchors support the bridge. Because it does not have a stiffening truss for support, the bridge swings as vehicles and pedestrians cross this stretch of the Levisa Fork. Some parts of the bridge, most notably the steel supports, were replaced with the widening and upgrading of U.S. 23.[68] A few historic bridges, such as the Pauley Bridge on U.S. 23 just north of Pikeville, made use of these same designs but incorporated native stone in constructing the towers. The Pauley Bridge is listed on the National Register of Historic Places. There are many examples of historic highway bridges as well. In Prestonsburg, two reinforced concrete arch bridges span the Levisa Fork and have also been added to the National Register of Historic Places.

Footbridge over creek in an unidentified coal town in Harlan County, Kentucky, circa 1940. Works Progress Administration Collection, Public Records Division, Kentucky Department for Libraries and Archives.

Railroads

Railroads served as the lifeline of the Big Sandy River Valley during the first part of the twentieth century. Four major railroads, along with countless branch lines, hauled thousands of tons of coal out of the region everyday. The railroads brought the equipment, supplies, and people needed to open mines. Although coal trains still rumble through the valley today, fifty years ago they were everywhere, loading coal at the tipples, shuttling empty coal cars back to the mines, bringing in consumer goods, and transporting people around the region. Railroad tracks wound through all of the towns, sometimes right down the middle of the main street. Residents became accustomed to the constant roll and rattle of slow-moving coal trains running between the mines and railroad yards day and night. Massive steam engines pulled the heavy loads through the mountains but also produced noise, smoke, pollutants, dirt, and ash. By the 1950s, steam engines were replaced with diesel locomotives, trucks, and automobiles. As the coal economy declined, so did the railroads.

Although these majestic steam engines and many of the historic railroad structures are gone, there are still constant reminders of how important the railroads were in the coal industry. Visitors to the Big Sandy region will encounter many railroad crossings, bridges, tunnels, and underpasses. There are old railroad yards, train depots, and countless miles of abandoned track. The old railroad grades can provide important clues about the layout of company coal towns. In the late twentieth century, the valley has attracted a small but growing number of rail fans who have recorded railroad history, helped restore cabooses, coal hoppers, and locomotives, and developed the railroading heritage of the region as a tourist attraction.

Locomotives, Railroad Cars, and Cabooses

There are only a few historic steam locomotives, railroad cars, and cabooses that remain in the coal towns of the Big Sandy River Valley. In the southern Appalachian mountains, only the most powerful engines were useful when pulling long trains of coal hoppers up steep mountain grades. One particular type commonly seen in historic photos of the region was the Mallet locomotive. By the 1950s, diesel engines replaced steam. The Norfolk & Western, the last major railroad in the country to make the switch, ran their final steam locomotives through the Big Sandy River Valley in 1960. There are only a few vintage steam engines in the region, but plans are

Top: Passenger and coal train at Williamson, West Virginia, 1953. Norfolk & Western Historical Photograph Collection, Virginia Polytechnic Institute and State Universities Libraries. Bottom: Loaded coal hoppers in the Williamson, West Virginia, yard. Eastern Regional Coal Archives.

under way to bring more back as railroad heritage projects progress. Some locomotives, along with passenger cars and cabooses, have remained in coal mining towns and commercial centers. A few have been refurbished and remodeled and now serve as tourist information centers and museums. Others have become part of the local landscape, placed in parks or recreational areas. Most passenger cars, cabooses, and freight cars were sided with wood until the 1920s when metal was more commonly used.

Railroad Yards

Trains brought coal hoppers from the mines to regional marshaling yards where the cars were sorted. When enough loaded coal cars were sorted they were joined together and placed on side tracks until a complete coal train could be assembled. A road train took the assembled hoppers along the main line to a classification yard where the hoppers were further sorted for their final destination.

In the Big Sandy region, the CSX system had several marshaling yards, located at Elkhorn City, Martin, and Russell, Kentucky. The Norfolk-Southern had two large yards at Williamson and Bluefield, West Virginia. The classification yard at Russell, Kentucky, was once regarded as the largest yard facility in the world. These yards were large, with many parallel tracks for sorting, storing, and placing coal hoppers. They were the site of constant traffic and are still active, on a much smaller scale, today. Smaller yards were located at Martin, Elkhorn City, and Shelby, Kentucky, and Eckman, Kimball, and Gilbert, West Virginia. In the Big Stone Gap area, the Norton yard remains an important transfer site. There were many structures and facilities located at marshaling or classification yards. Most had observation towers, offices, and engine repair shops. In larger yards locomotives were serviced in engine roundhouses, where massive turntables spun to place locomotives over repair pits or to redirect them to side tracks.

Coaling Stations and Water Tanks

Railroads were once the main consumers of coal, using between one-fifth and one-fourth of all of the coal produced in the United States. Freight trains pulling very heavy loads used much more coal than passenger trains, and in mountainous areas they consumed even more fuel. Steam locomotives used coal to produce power and had to be regularly refueled during runs through the Big Sandy region. Locomotive firemen continually shoveled coal and fed water into the engine, and the heat generated by the burning coal turned water into steam, powering the engines. Later, mechanical stokers did this grueling work.

At a coaling station, coal was dumped into the tender of the locomotive. Wooden coaling stations, common along tracks in the early twentieth century, resembled small tipples. A large holding tank was located above or alongside the tracks, and rectangular chutes positioned over the locomotive filled the tender. By the 1930s most coaling stations were made of steel and concrete. Along with coal, locomotives had to regularly replenish their supplies of water. Water tanks were located at rail yards, depots, and tipples to provide fresh water. When the locomotive stopped at a water tank, the fireman brought a spout into position and filled the tender with water from the tank. For efficiency and convenience, coaling stations and water tanks were often located side by side. The switch to diesel fuel in the 1950s ended the need for coaling stations and water tanks and many were demolished. Today there are few remaining coaling stations and water tanks in the Big Sandy region.

Railroad Stations

Most mining towns had railroad stations. Even the most remote mining towns of the Big Sandy region were connected to the three major railroads, the Norfolk & Western, Chesapeake & Ohio, and the Louisville & Nashville. Regional commercial centers and most company towns had depots that transferred freight and offered passenger service to nearby towns and cities. Even small coal camps usually had a platform or shelter alongside the tracks. Railroads served as the lifeline of many mining communities, and until the mid-twentieth century were often the only connection to the outside world. Passenger service allowed people to travel to nearby towns to shop, enjoy local entertainment, or visit friends and relatives. Railroad depots also handled freight. Trains brought small packages, mail, catalog orders, and supplies for local merchants. A typical coal town depot received a great variety of goods that eventually made their way to the shelves of the company store. Basic mining equipment and supplies, foodstuffs, and exotic goods such as fruits and vegetables were common arrivals at the depot.[69]

There were usually only two or three men who handled all the responsibilities of the depot. They sold tickets, announced departures and arrivals, signaled incoming trains, processed and delivered freight, and operated the local telegraph station. The depot was an exciting place. Trains brought news and information and new people to the town. In company towns, detectives often monitored the arrival and departure of passengers, especially if they were suspected of union organizing activities. A typical depot was a long, rectangular, single-story building parallel to the tracks. One end of the building served the needs of passengers. There were waiting areas, restrooms, and often vendors. A coal-burning stove warmed the area in winter months. Another part of the depot was for freight, the shipping and receiving of packages. On the platform were benches, wagons and carts for baggage, and mail sacks. The railroads set a new highly accurate standard for time, so locals set their watches by the clock at the local depot.[70]

These railroad depot buildings can still be found throughout the region. In many cases the tracks were removed long ago, but the depot building remains. The location of the depot was usually central, very close to stores and company offices. In a number of communities depots have become popular sites for small museums or historical societies. In Jenkins, the depot building now houses the David A. Zegeer Coal and Railroad Museum dedicated to the coal mining heritage of the Jenkins-McRoberts area. In Pikeville, the former Chesapeake & Ohio Depot has been placed on the National Register of Historic Places and has been renovated as city office space. The depot at Bramwell, West Virginia, has been rebuilt to serve as a tourist information center.

The former railroad depot at Jenkins, Kentucky, now serves as the David A. Zegeer Coal and Railroad Museum.

Railroads opened the valley to development and were often the only connection coal towns had to the outside world. Today abandoned railroad tracks, which may now appear to be just overgrown paths, can guide visitors to the coal towns of the Big Sandy region. Even in areas that have long since been abandoned, there are usually clues along the tracks that show how these towns were once laid out, and, with a bit of imagination, one can picture the town during its heyday. Along abandoned tracks there can be found the foundations of company buildings, mining equipment, and housing. Railroad tracks can lead to abandoned mine sites that have long since been sealed or obscured by overgrowth.

Other Buildings

A visit to a typical surviving coal town of the Big Sandy River Valley will reveal a variety of buildings and building styles in addition to mining structures: company office buildings, hospitals, clinics, city halls, fire stations, independent commercial structures, theaters, service stations, waterworks, and stores. Because sandstone and limestone are common in the Big Sandy

Once the company store and office at Itmann, West Virginia, this native stone building was placed on the National Register of Historic Places in 1990.

River Valley, many early-twentieth-century structures were made of native stone. Immigrant stoneworkers, especially Italians, built many of these structures from stone quarried locally. Lynch, Kentucky, built by U.S. Steel in 1918, has many fine examples, including a city hall, bathhouse, company offices, and fire station. Benham, Kentucky, built by International Harvester in the 1920s, has several excellent examples, and Jenkins, Wheelwright, and Van Lear also have a few. During the Great Depression, labor intensive projects carried out by the New Deal's PWA and WPA temporarily revived the use of native stone in building construction. Many of these buildings were schools, some of which remained in use into the 1990s.[71]

Cemeteries

Cemeteries can often provide vital information about mining communities. While most southern Appalachian communities have church or family based cemeteries, many company towns provided cemeteries for their employees and town residents. Cemeteries were usually segregated, with separate burial areas for white and black residents. There were often sections for immigrants as well. Tombstones can often provide excellent demographic information about the ethnic makeup of a community. With this basic overview of the landscape of the region, its history, and an understanding of the sites and structures found in these communities, visitors can begin their exploration of the Big Sandy River Valley and many historic coal towns.

The cemetery at Hellier, Kentucky, reveals the ethnic makeup and social divisions of the community.

Up the Tug Fork through the Coal Towns of West Virginia

4

Many of the Big Sandy River Valley coal towns in West Virginia lie along the path of the Norfolk & Western Railway, now paralleled by U.S. 52. This route can be reached from Interstate 64 just west of Huntington, West Virginia (pop. 51,502), at Kenova. It winds south through Wayne County, meets U.S. 119 at Williamson, and continues up the Tug Fork. U.S. 52 allows access to many historic coal towns in Mingo and McDowell Counties, as well as the county seats of Williamson and Welch. In McDowell County, U.S. 52 has been designated as part of the Coal Heritage Trail, a heritage tourism route that passes through the southwestern part of the state. Beyond Welch, where the headwaters of the Tug Fork originate, the highway continues into Mercer County. For travelers continuing south out of the Big Sandy River Valley, U.S. 52 leads to Bluefield, an important railroad town, and toward the mining district of Pocahontas, Virginia, a great place to experience the area's coal mining heritage.

Before venturing into coal country, rail fans may want to stop at the Huntington Railroad Museum, opened seasonally and operated by the Collis P. Huntington Railroad Historical Society. It is located in Ritter Park, at Memorial Boulevard and Fourteenth Street in Huntington, and is one of the few spots in the Big Sandy region to see a variety of historic railroading equipment and vehicles. A Chesapeake & Ohio Mallet Freight Locomotive, a design used from 1914 to 1948 for hauling coal on rough track, is on display. The park features cabooses, wood scooters, and a flat car. A wooden framed handcar, used in John Sayles's 1987 movie *Matewan*, can be pumped down an eighty-foot track. Passenger cars are being restored at the South Yard at 1101 Eighth Avenue and Eleventh Street. At

Heritage Station, near the Ohio River levee, there are railroad theme restaurants and shops and several locomotives and boxcars on display.

As U.S. 52 heads south from the Huntington area, it winds through rolling farmland and then begins to enter the mountains south of Fort Gay. Wayne County (pop. 42,903) has had very little coal mining,[1] and although there were some small coal towns in the early twentieth century, visitors will find surprisingly little of the coal mining past until they cross into Mingo County north of Kermit.[2] There is, however, constant coal truck and railroad traffic along this stretch of U.S. 52.

The Tug Valley of West Virginia was lightly populated and primarily agricultural until the late nineteenth century. It was the home of the Hatfield clan, and many reunions and reenactments continue to celebrate the feud. As the Ohio Extension of the Norfolk & Western Railway moved north in the 1880s, there was new attention paid to the valley's vast timber and coal resources. As settlement increased, Mingo County (pop. 28,253) was carved out of Logan County in 1895 and named for an area Indian tribe. Around 1900 the Norfolk & Western Railway opened up the Williamson coalfields to markets in Ohio and Virginia. By 1910 there were thousands of miners working in Mingo County.

U.S. 52

Borderland

U.S. 52 and U.S. 119 merge just south of Maher, West Virginia. The first site worth noting along this route is Borderland (pop. 100), a rapidly disappearing community soon destined to be a ghost town. It is located six miles north of Williamson. Borderland has an interesting history, but little of the town remains today. The widening and rerouting of U.S. 52/119 has cut the town in two and destroyed many of its buildings. Further flood control work will eventually eliminate the remaining structures along the Tug Fork. Most of the mining equipment and sites along the river have been demolished and only a handful of foundations remain.

Borderland was a company town built by the Borderland Coal Company after 1904. The mines were located on the Kentucky side of the Tug Fork and coal was moved on a tramway across the river to a tipple at Borderland, West Virginia, where it was loaded into coal hoppers. Borderland was spread over both sides of the river, with company offices and buildings in West Virginia and miners' houses on the Kentucky side. By the 1920s there were more than ninety homes, many made of brick, in Borderland. On the Kentucky side there was also a school, drum house, upper tipple, and machine shops. Supervisors' homes, a company store, the engine house, a powerhouse, and the lower tipple were located in West Virginia. Borderland prospered and made record profits during World War I. It anticipated further growth and invested in a new tipple in 1918, but the company had financial problems and barely survived the 1920s decline in coal prices.[3]

Borderland Coal was dominated by its first president and main stockholder, Edward L. Stone; his life was a classic rags-to-riches story of the late nineteenth century. He worked his way up through a Roanoke, Virginia, printing firm, became manager, and expanded the company's markets. After the death of the owner, he gained control of the company and married into a prominent local family. Stone eventually invested in banks, iron companies, foundries, and Borderland Coal. Borderland was a typical company town, providing only a few services or facilities to workers and their families. Although Stone was very generous with charities and civic organizations, Borderland Coal provided little for its employees.[4]

Borderland Coal was also vehemently anti-union. During the West Virginia Mine War, the company used every tactic available to destroy the influence of the UMWA. Baldwin-Felts guards harassed union organizers, sympathizers were fired and evicted, and elaborate espionage systems were developed to identify anyone involved in union activities. In the 1910s, Borderland joined other Mingo County coal operators in fighting the union, and tensions peaked in 1920 following the Matewan Massacre. Because the town and the mining operations straddled the Tug Fork, Borderland was frequently the site of violence. Union men became involved in a guerrilla war with Baldwin-Felts men and state police who protected the town. Guards were attacked and strikebreakers were harassed. Because of its position on the river, snipers from West Virginia fired on nonunion workers in Kentucky. Company guards returned fire. In November 1920 there was a series of shootings at the mines, houses of nonunion workers were burned, and the engine room was dynamited. During the worst phase of the fighting, federal troops were strategically stationed at Borderland.[5]

Like the rest of Mingo County's coal operators in the 1920s, Borderland Coal prevailed and defeated the union. But this did little to solve its growing financial problems. Declining coal prices and overinvestment resulted in a crisis by the late 1920s. In 1932 the company filed bankruptcy and closed its mines two years later. The town of Borderland barely survived the Great Depression. The surrounding area was later mined by several coal companies in the 1940s and 1950s, but the town never recovered.

Borderland is no longer clearly marked, but there are two small clusters of houses, one on each side of the highway. The Deskins Drive and Borderland Road exit off U.S. 52/119 marks the former boundaries of the town. West of U.S. 52/119 there are approximately ten houses remaining with a prominent landmark, the Borderland Baptist Church, a modern brick structure. The houses are one-story, three-bay brick homes with clipped gables and front porches. In the thickly wooded area alongside the river is the site of the tramway that once brought coal from the Kentucky mines. East of U.S. 119 are a few remaining houses from the early twentieth century and some foundations located about one-half mile from the highway. The Kentucky section of Borderland is not accessible from West Virginia, and the remains of the tipple and tramway were

Edward Stone in his Virginia study, circa 1914. Borderland Coal Company Papers (#382), Albert and Shirley Small Special Collections Library, University of Virginia Library.

Damaged tram at Borderland, circa 1916. Borderland Coal Company Papers (#382), Albert and Shirley Small Special Collections Library, University of Virginia Library.

demolished in the 1990s. Directly across the Tug Fork, on the Kentucky side of the river, there are scattered foundations of houses but little further evidence of Borderland.

Williamson

About six miles south along U.S. 52/119 is Williamson (pop. 3,300), which has served as a transportation and regional commercial center for the Tug Valley for more than one hundred years. It has several interesting sites and serves as a gateway to many area coal towns, especially south along U.S. 52. Williamson is also where U.S. 119 crosses the Tug Fork and continues west into Pike County. A quick trip across the river and up U.S. 119 leads to Stone, a major Tug Fork coal town in the early twentieth century. Continuing west, travelers can follow U.S. 119 into Letcher and Harlan Counties. Heading east on U.S. 119, travelers leave the Big Sandy River Valley but can visit two additional important sites. The early model company town of Holden lies about twenty-eight miles to the east, and Logan, a regional commercial center, another five miles farther. Leaving U.S. 119 at Logan and following WV 17 east leads to Blair, the site of the Battle of Blair Mountain.

The Williamson area was developed in the 1890s as the Norfolk & Western Railway penetrated the region and opened area coalfields. The town was named for Wallace J. Williamson, who operated a farm in the area now covered by the city and sold his land to the railroad. It was incorporated in 1892, became the Mingo County seat in 1896, and a city in 1905. The town lies in the center of the "Billion Dollar Coal Field," a part of the Williamson coalfield, an area with more than one hundred mines operating by the 1920s.[6] Williamson was a major city of more than thirty-five hundred residents by the early years of the century, and its population almost tripled over the next thirty-five years. Despite recent decline, Williamson still serves as an important regional commercial center and is the location of a major railroad classification yard that dissects the town and provides the main means of transporting coal north and east. A 1990s flood control project now protects the city from the raging waters of the Tug Fork.

Downtown Williamson has a few important sites related to the coal industry. The railroad is the largest single employer in Williamson, and the local facilities make up the largest coal marshaling yard in West Virginia. The residential areas of the city are built on the hillsides surrounding the town and offer good views of railroad operations and the surrounding Tug Valley.

The Tug Valley Chamber of Commerce is housed in one of Williamson's unique buildings, the Coal House, located in the courthouse square. Listed on the National Register of Historic Places, the Coal House is a one-story structure with a large, arched front entrance and tall narrow windows. Sixty-five tons of local coal form the four outer walls. The idea to construct a coal house was that of O. W. Evans, the manager of the Norfolk & Western Railway's Fuel Department. In 1933 he convinced local officials to help promote the coal industry by making a building out of coal. Materials, labor, and cash were donated by local coal operators and merchants. The Coal House is frequently sealed and varnished to maintain its shiny black outer appearance and is fully insured. It does not represent any fire hazard. Until the 1950s, coal operators used the Coal House as a way to publicize the importance of coal in the local economy. Today the Coal House is proudly occupied by the chamber and continues to promote the coal industry in Mingo County, as well as area tourism. The chamber provides maps and walking tour brochures of some of the historic structures in Williamson. It also supplies information about surrounding attractions,

Top: The Coal House in downtown Williamson, West Virginia. Bottom: Williamson's rail yard contains more than one hundred miles of track.

sites, and lodging. The brief walking tour features a variety of historic buildings from the early twentieth century, including churches, commercial buildings, and residences. Examples of architectural styles in the commercial core of the city include Victorian, Italianate, and art deco. The main residential areas highlight styles ranging from simple workers' housing to elaborate Victorian and classical revival homes. The old Williamson cemetery is located on Reservation Hill and is the final resting place of the city founders and many early residents including the Williamson family.

Williamson's rail yard dominates much of the city and can be seen especially well from U.S. 52 when traveling south. It contains more than one hundred miles of track, making it the largest marshaling yard in West Virginia. The railroad complex includes a roundhouse, swing bridge, and machine shops. The new Williamson Area Railroad Museum, temporarily housed on Second Avenue, is a good place to learn about railroading in the Tug Valley. The founders have been active in promoting model railroading events, organizing the Williamson Railroad Festival in October, and preparing exhibits that highlight the city's rich transportation and coal history. On Fourth Avenue the old Norfolk & Western passenger depot is now the Williamson City Hall.[7] From Williamson, travelers can turn east, cross west into Kentucky, or continue south along U.S. 52/ WV 49 to tour the coal country of the Big Sandy region.

East on U.S. 119: Beyond Mingo County

For travelers heading east on U.S. 119 out of the Big Sandy River Valley, there are many sites related to the great coal boom of the early twentieth century. Much of southern West Virginia has recently been designated as a National Coal Heritage Area and will undoubtedly see more of its coal mining history preserved, restored, and celebrated in coming years. Logan County (pop. 37,710) lies in the southwestern part of West Virginia, north of Mingo County on U.S. 119. Most of the county's hilly terrain is drained by the Guyandotte River in the north, placing it outside of the Big Sandy River Valley. The Logan coalfields were opened by the Chesapeake & Ohio Railroad. They were closely tied to the Big Sandy area and were the site of many events in the West Virginia Mine War of 1920–21. A brief trip east on U.S. 119 leads to the model company town of Holden, the commercial center of Logan, and the site of the Battle of Blair Mountain.

During most of the nineteenth century, Logan County was a lightly populated farming area with a limited timber trade. Even though there were known deposits of coal, they were worthless without a way to ship them to markets. Development was extremely slow and was confined to the western edge of the county. Flatboats floated down the Guyandotte River to the Ohio, providing the only transportation link to the outside. In 1895 Logan County was split into two parts. The southwestern half became Mingo County.[8] After 1900 the Chesapeake & Ohio Railroad began to penetrate the county and reached the village of Logan in 1904. From there, it extended a series of spur lines into remote hollows where coal mining was soon developed. The mining industry brought big increases in the county's population, and new coal towns quickly appeared. From 1900 to 1920 the population of Logan County increased from about seven thousand to more than forty-one thousand people.

The development of the Logan coalfields continued into the 1950s when the county was home to more than seventy-seven thousand residents. Declining coal prices and new mining methods soon led to a sharp decline in employment, and out-migration took a heavy toll. By the 1960s many of the coal towns in Logan County were abandoned. Logan County now produces record amounts of coal, but much of it is strip-mined. The controversial practice of mountaintop removal threatens not only the landscape but also many of the historic mining communities described in this guidebook. Although mining is still an important industry, Logan County has experienced some limited diversification in its economy with the expansion of retail, service, medical, and educational facilities. As U.S. 119 is upgraded to a major four-lane highway, Logan County may continue to attract development and preserve more of its coal mining heritage.

Holden

The model company coal town of Holden (pop. 300) lies on U.S. 119 and was developed by the U.S. Oil Company in 1902. Two men who had large interests in U.S. Oil, Colonel William H. Coolidge and Albert F. Holden, surveyed the property and saw great potential for the Logan coalfields. Coolidge was a Boston financier who was convinced that the thirty thousand acres they were surveying could yield vast profits. Holden was an engineer by training and recognized the high quality of coal in the area.

At the time, there were only a few scattered homes in the area and a handful of small coal mines and timber operations. The Chesapeake & Ohio Railroad was still forty miles away. After the men acquired the property, the Island Creek Railroad was constructed to connect Logan County with the Chesapeake & Ohio.[9]

Albert Holden was determined to make the new mining operation and company town a model community in the coalfields. U.S. Oil hired local laborers to begin the construction of the town and its mines. Local stone was quarried, timber was cut for buildings, and streets were laid out. Because there were few local residents, a large town was built and workers were recruited from outside of the area. Coolidge, later testifying before a U.S. Senate committee, stated that they intended to make the town the best anywhere. He noted that the houses built were "better than anybody else's." Coolidge was not shy about his motives; he and Holden wanted a good environment for their employees but also hoped to keep the UMWA out. He took responsibility for planning the town and insisted that it be a good place to live. He arranged for well-stocked company stores, a recreation center, and social activities. He was able to convince the company that this was an investment as well as an expense. The company soon promoted the new town as a model mining community, named it for its chief planners, Albert Holden. Holden then became president of the U.S. Coal and Oil Company, serving until his death in 1913.[10]

The mining operations at Holden were extensive. The first shipments of coal left the mines in 1904, and after only six months the company had mined more than one hundred thousand tons. By 1910 there were three mines opened at Holden. U.S. Oil was able to negotiate reduced rates on rail and river transportation by threatening to build their own railroad north to Huntington, competing with the Chesapeake & Ohio. Coal was mined at Holden from 1902 to 1915 for the U.S. Coal and Oil Company. A subsidiary company, Island Creek Coal Sales, marketed its coal. After 1915, U.S. Coal and Oil was reorganized and became Island Creek Coal Company, one of the leading developers of coal in West Virginia and east Kentucky throughout the twentieth century.

The operations at Holden thrived, and by the 1920s there were twelve mines in the area, along with machine shops, tipples, managers' houses, and a company hospital. Workers enjoyed a clubhouse, community churches, libraries, bowling alleys, company stores, a theater, and an opera house.[11] Holden was credited with raising the standard of living in area

Holden's main street was once part of U.S. 119.

coal camps as companies competed for laborers. The 1920s brought further expansion in the mountains of Logan County by Island Creek and other companies. By 1925 there were fifty companies and more than one hundred mines operating in these coalfields.[12]

The Great Depression brought problems to the coal industry, but Island Creek began an extensive program of modernization and emerged as a solid, profitable company by the 1940s. The company invested heavily in its mines and in 1933–34 the Holden operations were expanded. A new tipple with twelve loading points and an extensive conveyor system was developed to bring coal from the mines.[13] Unfortunately, the Holden mines were also the site of a tragedy. In 1943 an explosion created a burst of gas that killed eighteen men at Island Creek mine No. 22. The blast hindered rescue attempts, and it took nine days to reach the bodies. Regular coal mining continued in the area until the 1960s, and Holden served as the headquarters for Island Creek's mining operations in Logan County until the 1990s.

Holden lies alongside U.S. 119, now a four-lane highway that has bypassed the town. Old U.S. 119 serves as the exit into the town and its main street. Holden has many buildings remaining from the early twentieth century and has only recently suffered with the relocation of the highway. In the center of Holden is the former company store, company headquarters, and church. The company office building, a three-story brick and concrete building in the center of the town, is still in use. About sixty to seventy houses remain in the town. Most are two-story side-gabled wood frame houses with front porches. Because U.S. 119 ran directly through the center of Holden until quite recently, and carried a considerable amount of traffic, it was widened as much as possible, leaving very narrow sidewalks and many front porches right up against the road. The railroad tracks cut directly through the town and lead to the former mine sites.

Logan

Four miles east of Holden is Logan (pop. 1,630), the regional commercial center and political seat of Logan County. It was a small trading town for farmers in the 1800s but grew as timber and coal businesses developed along the river. By the 1910s, the downtown commercial district was a lively place as thousand of miners and their families rode the rails into town on weekends. Merchants opened stores in the lower levels and often lived with their families on the second floor. Logan had its share of theaters, salons, barbershops, pool halls, and recreation centers as well. The outward appearance of the central commercial district has not changed a great deal in the past one hundred years. Logan was the center of action during the last phase of the West Virginia Mine War when miners threatened to march on the town and attack Sheriff Don Chafin, who had an army of mine guards, deputies, militia volunteers, and detectives.

The Women's Club of Logan Library, at 581 Main Street, was once the home of Logan County's notorious sheriff Don Chafin. Chafin was a prosperous, hated, and feared man whose corrupt, often brutal rule and close affiliation with the county's coal companies earned him the reputation as the "Czar of Logan." His beautiful home on Main Street stands in sharp contrast to those of most county sheriffs in the area and is one few historic structures remaining from the West Virginia Mine War. On an elevated lot and bordered by a wall of cut stone and concrete, Chafin's home is surrounded by nicely landscaped grounds with dogwood and evergreen trees. The large, two-story house with bay windows, a stone chimney, and several dormers was built around 1900. Its architecture is a blend of Victorian styles and the simple functional designs found in many coal towns. In 1994 it was placed on the National Register of

Historic Places. One can only imagine the political and criminal intrigues planned in this home where Chafin lived until 1933. He moved from this residence to Huntington, where he died in 1954.[14]

Four miles north of Logan is Chief Logan State Park, located on thirty-three hundred acres of reclaimed coal mining land. Formerly the site of the Merrill Coal Company, the land was stripped of mining and railroad equipment and was developed as a recreational area in 1969. The Coal Mine Trail leads hikers through the park, where they can view old mine portals, storage silos, and railroad grades. Also on display is a 2700 class stream locomotive donated by the Chesapeake & Ohio Railroad. A Coal Miners' Memorial, dedicated to the "coal miners of West Virginia who perished from mine accidents and lung disease" is also located in the park.[15]

Blair

Beyond Logan, the town of Blair (pop. 200) is located near the base of Blair Mountain on WV 17. Two years of conflict between coal operators and union miners in West Virginia culminated in September 1921 with the Battle of Blair Mountain. A quiet community of a few hundred residents with a few houses and a post office, Blair now shows little evidence of having been the scene of such a tragic event in American labor history. A West Virginia historical highway marker indicates the site of the battle, located

Some of the few remaining homes at Blair, West Virginia.

on WV 17 about eight miles east of Logan, between Ethel and Blair. The town of Blair now faces a problem far worse than neglect: it may soon disappear completely. Blair is being threatened by a massive mining operation on Blair Mountain that is using the technique of mountaintop removal to take up to 300 feet off the top of the 840-foot peak. Arch Coal is using new dragline equipment to shears layers off the top of Blair Mountain, exposing multiple seams of coal below. Draglines can cover an area almost twenty stories high and weigh up to eight million pounds. Once the topsoil is removed from the mountain, rock is blasted away exposing the coal seam. Coal is then shattered into pieces and hauled away by truck. The historic town of Blair, located at the base of the mountain, has suffered greatly from the constant blasting, debris, and coal dust. Through a subsidiary, Arch Coal purchased many of the houses in Blair, which are now vacant. The battle for Blair has moved to the courts as remaining residents fight to save their homes and community.[16]

Up the Tug Fork to McDowell County on U.S. 52/WV 49

To enter the heart of West Virginia's coal country, follow U.S. 52/WV 49 south through Mingo, McDowell, and Mercer Counties. About four miles south of Williamson, a West Virginia historical highway marker indicates the spot where sections of the Ohio extension of the Norfolk & Western Railroad were joined in 1892. The project involved five thousand laborers and took two years to complete. Once connected, the Norfolk & Western became a major Atlantic-Midwest carrier and the main outlet for the Pocahontas coalfields.

Matewan

Matewan (pop. 498) is located on the Tug Fork, sixteen miles south of Williamson on WV 49. It was founded in 1892 with the arrival of the Norfolk & Western Railway, incorporated in 1895, and prospered as a railroad junction and regional commercial center. The Tug Valley around Matewan is where the Hatfield family resided and where several major events in the feud took place. Matewan was not a company-owned town, but it was surrounded by small coal towns. Its growth was directly connected to the mining industry and the Norfolk & Western Railway.

Matewan served the many coal camps of the Williamson coalfields and was the site of several important events in the battle to organize coal miners in southern West Virginia. One-quarter mile up Mate Creek was the site of the Stone Mountain Coal Company, the scene of numerous incidents during the West Virginia Mine War, vividly portrayed in the 1987 John Sayles film *Matewan*. The film, which stars Chris Cooper, James Earl Jones, Mary McDonnell, and Will Oldham, depicts the conflict between coal operators and union organizers that culminated in the famous "Matewan Massacre." Even though the movie was not filmed in Matewan, it did create a renewed interest in the region, and a small but growing number of tourists now frequent the town. The Matewan Development Center has been instrumental in preserving and promoting area history, especially coal mining and the feud. The listing of the Matewan Historic District on the National Register of Historic Places in 1993 and the designation of the Matewan Historic District as a National Historic Landmark by the U.S. Department of the Interior in 1997 has made the town a center of many new heritage tourism projects. It has become one of the first places in southern West Virginia to truly capitalize on its coal mining past.

Matewan has a number of sites related to the mine war and a visitors' center that actively promotes and preserves the town's heritage. Despite many floods and several fires, the town's main thoroughfare, Mate Street, appears much as it did in the early twentieth century. One of the unique features of Matewan's commercial district is that the buildings have two facades, one facing the railroad tracks and one facing Mate Street. The Norfolk & Western tracks were originally the town's main thoroughfare. The front of these commercial buildings, with recessed storefronts and display windows, faced the tracks. On the other side of the buildings are mirror image storefronts that face Mate Street and serve as the main entrances today.

The Matewan Development Center is located in the G. W. Hatfield Building, a large three-story commercial structure with four bay windows, built in 1911 by Mingo County politician Greenway Hatfield. The center features exhibits on the Hatfield-McCoy feud, the Matewan Massacre, and the coal mining history of Mingo County. It also organizes local events that promote tourism and has undertaken an extensive oral history project to record the experiences and recollections of area residents. The center provides maps and information about Hatfield-McCoy sites and the Coal Heritage Trail. Hatfield-McCoy events have become quite popular in recent

The site of the Matewan Massacre.

years and a Hatfield-McCoy celebration now takes place along the Tug Fork each year in the late spring. The Hatfield cemetery is located on WV 44, in nearby Logan County, south of Stirrat. A West Virginia historical highway marker notes the site, and a life-sized statue of William Anderson "Devil Anse" Hatfield towers over the graves (see Pikeville, Kentucky). There is a driving tour map available, and many of the sites have been documented and appear on the National Register of Historic Places.

A Matewan walking-tour brochure identifies buildings along Mate Street. Most are typical of commercial structures found in early-twentieth-century coalfields. They are two-story brick or stone structures with simple windows and some pressed metal ornamentation. A recently completed flood control project left the historic commercial center protected and intact. The new floodwall, completed in 1996, is more than two thousand feet long and features a series of illustrated panels that depict the history of Matewan and the Tug Valley.

The Norfolk-Southern Railroad tracks wind through town behind the buildings of the commercial district. The train depot (demolished in the 1960s) stood near the John Nenni Building, which was originally

Sid Hatfield and Mayor Testerman lie in Buskirk Cemetery, across the Tug Fork.

three separate buildings—the Testerman jewelry store, a restaurant, and a department store—when constructed in 1910. The Testerman store was located in the large, three-story commercial structure. Mayor Testerman and his young wife, Jessie, operated the store and sold jewelry, tobacco, small gadgets, and instruments. The two-story segment of the Nenni Building was the site of Baldwin-Felts informant Charles E. Lively's restaurant. This was the site of many UMWA meetings and where Sid Hatfield divulged many of the details of the Matewan Massacre to Lively.[17] The Buskirk Building, located at Hatfield Street and Mate Street, was the site of the Urias Hotel, the headquarters for the Baldwin-Felts detectives. The two-story brick Matewan National Bank and Old Post Office Building was constructed in two sections. The section facing Mate Street was the site of the bank, now restored to its original appearance and featuring round-headed windows on the first floor. The section facing the railroad tracks with a corbeled parapet and flat-headed windows was once the post office. A West Virginia highway marker indicates the site of the Matewan Massacre, in the alley opposite the post office. Bullet holes are still visible in some of the surrounding buildings.

The one- and two-story houses along the west side of the road north of the commercial district in Warm Hollow and the Coleman Addition are typical of those built for miners and their families in the early years of the twentieth century. Stone Mountain Coal Company's settlement was east of town about one-quarter mile up Mate Creek near WV 49. Few structures remain today, and the site of the evictions is unmarked. Just south of Matewan is Thacker, a coal camp established in 1893 by the Thacker Coal and Coke Company. Today a few scattered houses along the railroad tracks are all that remain of the settlement.[18]

Several of the people involved in the West Virginia Mine War and the Matewan Massacre are buried in the Hatfield cemetery across the river in Buskirk, Kentucky. The cemetery lies alongside a road that leads up the hill to the radio tower, off KY 1056. Overlooking the valley are the graves of Sid Hatfield, Mayor Cabell Testerman, and Ed Chambers.

Red Jacket

Red Jacket (pop. 728) was constructed by the Red Jacket Consolidated Coal and Coke Company in 1905 and expanded during the coal boom of the 1910s.[19] Red Jacket Consolidated Coal and Coke was the largest coal operator in Mingo County, owning more than eleven thousand acres of land along the Tug Fork by the 1910s. Red Jacket miners were paid fairly well with an average annual income of about thirteen hundred dollars in 1920. The company provided good housing and services for their workers. More than one thousand residents had access to company stores, a company movie theater, and two schools run by the company, one for white children and one for black children. Most of the miners were native white and African American, but Red Jacket also had a small immigrant population of Hungarians, Spaniards, and Italians. Religion was encouraged by the company, and several itinerant preachers served the small church in town.[20]

As one of the largest employers in the Williamson coalfields, Red Jacket was particularly concerned about union activities and took steps to prevent organizers from influencing their workers. Red Jacket required its miners to sign yellow-dog contracts that forbade any employee to join a labor union, associate with any union organizers, or assist in any union organizing activities. Under the contract, the employee agreed "that he will not belong to, or affiliate in any way with" any union and "will not knowingly work in or about any mine where a member of such organization is employed." The company expressly notified any of its "EMPLOYEES WHO AFFILIATE THEMSELVES WITH such an . . . organization to IMMEDIATELY sever their connection with the company and to MOVE OFF THE PREMISES."[21]

Despite such warnings, in 1920 the UMWA launched a major effort to organize the southern West Virginia coalfields. In Matewan the union held meetings targeting Red Jacket miners, pressuring them not to sign the contracts. In September the Red Jacket Consolidated Coal and Coke Company sued for an injunction against the UMWA claiming violation of its contract right. The U.S. District Court agreed and issued an injunction against the union, and 315 coal operators in southern West Virginia joined in the suit. The court forbade the UMWA from in any way "molesting or interfering with or attempting to molest or interfere with the employees of the plaintiffs in the performance and fulfillment of their contracts" and from "entering upon the grounds and premises of the plaintiffs, or their mines to persuade them to join the United Mine Workers of America."[22] Other coal operators followed with similar requests. In one instance, the Pond Creek Coal Company was able to prevent organizers from informing incoming strikebreakers that a strike even existed. In 1923 the operators were granted a permanent injunction against UMWA organizing activities. The injunction was upheld on appeal in 1927 and the U.S. Supreme Court declined to hear the case.[23] These court actions were very effective and allowed Red Jacket to prevent the organization of its workers for the rest of the decade.

Red Jacket Coal survived the Great Depression and prospered during the 1940s, but by the 1950s the company scaled back its operations at Red Jacket. In 1956 they sold their properties and ceased mining. Other coal companies leased lands at Red Jacket and operated the mines. Island Creek was the last to mine coal here in the 1970s. The Georgia-Pacific Corporation currently owns much of the lands around Red Jacket and has numerous mining and timber operations in place.[24]

Red Jacket is located four miles east of Matewan off WV 65. The main road into town parallels the railroad tracks and houses built by the company in the early years of the twentieth century. Many of the homes are of the two-story L design with end-facing gable roofs and attached front porches. Others are modified versions of the two-story basic I design with end-facing gables and front porches. The old Norfolk & Western depot is located on the right after the railroad crossing. The one-story brick structure has been converted into a private residence. The road alongside the depot, Mitchell Branch Hollow Road, leads to the site of one of Red Jacket's main mines. The sealed entrance is on the right, about three-quarters of a mile up, behind a cluster of single-wide mobile homes.[25] WV 65 continues beyond the railroad tracks and passes a two-story frame building that was once the clubhouse, a two-story brick and block school, and a number of residences.[26]

Beyond Red Jacket, U.S. 52 continues south along the Tug Fork. The Norfolk-Southern tracks continue across Mingo County, winding through the valley along the river. Spurs connect several mining areas in Pike County, Kentucky, on the opposite bank. The Norfolk-Southern line passes through the small mining settlements of Edgarton, Cedar, Devon, Lindsey, and Glen Alum. Past Glen Alum, the tracks divide: one line follows the river into McDowell County and a second line runs northeast to Gilbert. Unfortunately, south of Red Jacket, U.S. 52 leaves the path of the rails and the river. It climbs through a particularly challenging area where Horsepen Mountain reaches an altitude of 2,524 feet. At Gilbert, where Horsepen Creek drains into the Guyandotte River, WV 80 joins U.S. 52 and the CSX line junctions with the Norfolk-Southern Railroad. From Gilbert, the road follows a twisting course south through the Guyandotte River Valley, part of Wyoming County, and into McDowell County. At

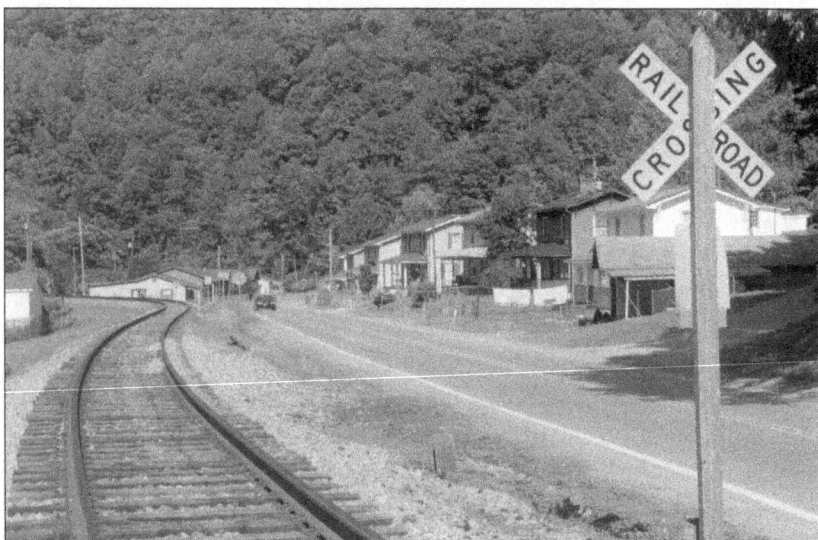

Houses and railroad tracks through Red Jacket, West Virginia.

Iaeger, which was once the western terminus of the electrified section of the Norfolk & Western, U.S. 52 once again parallels the rails and the river. From here all the way south to Bluefield, the tracks are always near the highway, the Tug Fork is close-by, and the many coal towns of the region can be easily reached by a short ride on state and local roads.

McDowell County (pop. 27,329) has been one of the major coal mining counties in West Virginia for more than one hundred years. Lying in the southernmost part of the state, McDowell County was created in 1858, carved from Tazewell County and named for James McDowell, governor of Virginia in the 1840s. Until the 1880s the county was one of the most remote and isolated parts of the state, with a small, largely self-sufficient population of about three thousand mountaineers. Only a small timber trade existed around the town of Welch. Many people were aware of the rich coal seams that ran through the county, but the lack of transportation prevented the development of a mining industry. The county was thrust into the industrial era in the 1880s when the Norfolk & Western Railway arrived. Production and population soared and the area experienced great growth into the mid-twentieth century.

McDowell County's coal came from the Flat Top–Pocahontas coal-field, an area of high grade coking coal. Much of the land in McDowell and Mercer Counties was developed by the Southwest Virginia Improvement Company, a subsidiary of the Norfolk & Western heavily financed by Pennsylvania investors. In the 1880s it opened the area to mining, brought the railroad through, and leased land to area coal operators. It later became the Flat Top Coal Land Association, the Pocahontas Coal and Coke Company, and finally the Pocahontas Land Company.[27] Mines were opened in Virginia in 1883, and a few years later coal mines were operating in McDowell County, financed by companies that bought and leased much of the land along the railway lines. As the railroad moved farther into the county, small company coal towns began to appear. By the early 1890s the county had twenty-five mines and almost one thousand coke ovens. The town of Bramwell and the regional commercial center of Welch developed as the headquarters for many of the coal operators. The mining boom drew people from the east, south, and other countries. McDowell County gained a reputation as one of the best places for southern blacks to seek employment and drew a large number by the turn of the century. By the 1910s the county had a population of almost 48,000, many of whom were African American and eastern

European. The increase continued into the 1940s, when the county's population peaked at almost 99,000. Today the county has fewer than 26,000 residents, but many historic coal towns remain within its boundaries.[28] U.S. 52 cuts through McDowell County, and beyond Welch it is an important part of West Virginia's Coal Heritage Trail. This part of the state offers visitors a tremendous variety of coal and railroad sites to explore. The Tug Fork has its origins as a major tributary of the Big Sandy River in McDowell County. It is fed by many creeks and streams in the southeastern part of the county. Travelers can cross the mountains and leave the Big Sandy Region, exploring nearby Bramwell, Bluefield, and Pocahontas, Virginia.

Welch

The seat of McDowell County is Welch (pop. 2,683), located on U.S. 52 about fifteen miles from Iaeger. It is situated where Elkhorn Creek meets the Tug Fork on a level area of the Elkhorn Valley. The central commercial district of Welch lies on the valley floor, while many of the residential areas are on higher ground on the surrounding hills. Welch was an isolated site along the river during most of the nineteenth century. There were no roads or bridges in this rugged part of McDowell County until 1880. The few people found in the area lived off the land and traded ginseng and furs to survive. The site was settled in 1885 by John Henry Hunt, who later sold the land to Capt. Isaac A. Welch, for whom the town is named. Captain Welch's accounts of great coal seams and timber in the surrounding hills persuaded Norfolk & Western's president Frederick J. Kimball to bring the railroad to Welch. There were no more than a dozen homes here when the first train arrived in 1891, but the next year Welch became the county seat and two years later was incorporated. As streets were laid out and people flocked to the town it became a thriving regional commercial center regularly drawing thousands of miners and their families who came to conduct business, shop, find entertainment, or socialize. The earliest buildings in Welch were made of wood, and a disastrous fire wiped out most of the central district in 1911. During the next few years the town was rebuilt using brick, stone, and masonry. Most of the buildings in the downtown area today were constructed between 1915 and 1930. The commercial district of Welch has been placed on the National Register of Historic Places.[29]

The regional commercial center of Welch, West Virginia.

For the first fifty years of the twentieth century, Welch served as the trade, banking, and entertainment center for people living in the surrounding coalfields. Saturday nights would find the streets jammed with mountaineers, African Americans, and Chinese, Japanese, and European immigrants. The congestion was sometimes compared to that of New York City, and visitors often commented on the cosmopolitan atmosphere of the community.[30] The downtown commercial district is typical of those found throughout the region in the early twentieth century. The buildings were built in the American commercial style, two- and three-story brick and stone structures with differing decorative elements. McDowell Street was the main commercial street filled with department and specialty stores, theaters, barbershops, and restaurants. Wyoming Street, which runs parallel to McDowell, is where most of the government buildings, banks, and offices were located.

The McDowell County Courthouse on Wyoming Street was designed by Frank Milburn in the Romanesque revival style, popularized by American architect H. H. Richardson. Built in 1893–94, this two-story

Sid Hatfield and Ed Chambers were killed here on the steps of the McDowell County Courthouse.

building with a square three-story tower was originally to be made of brick, but native stone was used instead. Although modern hardware and lighting have been added, and many minor renovations have taken place over the years, the interior retains much of its original appearance. An imposing cast-iron stairway leads to where the courtroom once filled the entire second floor, and a 1909 addition added much of the current office space. A West Virginia Historical Marker describing Welch is located in front of the building.[31]

During the West Virginia Mine War the McDowell County Courthouse was the scene of many hearings and trials, but it is best known as the place where Sid "Two-Gun" Hatfield and his companion Edward Chambers were assassinated by double agent Charles Lively and a group of coal company guards. The McDowell County Courthouse survives as a living artifact of the mine war and appears much as it did in 1921. It is listed on the National Register of Historic Places.[32]

Gary

McDowell County's premier company coal town, Gary (pop. 910), is located on WV 103 just south of Welch. The last major company town in the Big Sandy Region, surviving until the 1980s, Gary had a large, diverse population, underwent several periods of modernization, and has many structures remaining from the early twentieth century. According to many of its longtime residents, Gary was a very good place to live. Gary is also one of the best documented coal towns in this guidebook. Historian and former electrical engineer Mike Hornick has amassed a tremendous amount of material about the town and has written several summary histories. Dr. Stuart McGehee has also written about Gary and has arranged for many of the town's records to be transferred to the Eastern Regional Coal Archives, housed in the Craft Memorial Library in Bluefield.

The opening of the Gary coalfields was the result of a shrewd business deal by Bramwell banker Isaac T. Mann. Mann made an offer to purchase more than 200,000 acres from the Flat Top Land Association (a subsidiary of the Norfolk & Western Railway). The company attorneys jokingly offered the land to Mann for an option of $50,000. Mann promptly produced the money and arranged to return with the balance of the transaction. After making the deal, he headed directly to New York and met with John Pierpoint Morgan, who agreed to back Mann's venture. Mann then returned to West Virginia, sold the land back to Flat Top Land Association at nearly double the price ($20 million), and arranged for a lease of 50,000 acres for Morgan's newly formed U.S. Coal and Coke Company. Mann's profits allowed him to invest heavily in banking, land, and mining, maintain an estate in Bramwell, and serve as president of both the Bank of Bramwell and the Pocahontas Fuel Company.[33] His home is an important tourist attraction in Bramwell today.

The town of Gary was built by the U.S. Coal and Coke Company, a subsidiary of the newly formed U.S. Steel Corporation, in 1904. It was named for Judge Elbert H. Gary, who oversaw the reorganization of the Illinois Steel Corporation in Chicago and helped turn it into the modern, integrated Federal Steel Company. At the time, Andrew Carnegie, who had built the Carnegie Steel Company in the late nineteenth century, was planning to retire and was looking for someone to purchase his vast steel empire. Judge Gary helped persuade J. Pierpoint Morgan to purchase the company and combine it with Federal Steel. In 1901 these two giant steel producers merged with several others, forming the U.S. Steel Corporation. Judge Gary became chair of the board, and Charles M. Schwab served as the first president. In 1902 U.S. Coal and Coke, a subsidiary of U.S. Steel, was chartered and Thomas Lynch served as its first president.

The subsidiary leased a large tract of land in McDowell County and began preparations for a series of mining operations in what later became known as Gary Hollow. The Tug Fork branch of the Norfolk & Western Railway continued from Welch toward the site, opening up land for development along the river and its many tributaries, and over the next seven years twelve town sites were developed in the area. Like other areas of southern Appalachia, the local labor force was limited and often seasonal. Farmers looked at mining as a way to supplement their income during the off months. But Gary soon had a large labor force and actively recruited more as tracks were laid, mines opened, and new settlements developed. Many laborers were southern blacks fleeing the Jim Crow systems of southern states. Another large segment of the population was eastern European. Immigrants from countries such as Italy, Poland, Hungary, Russia, Czechoslovakia, and Greece arrived in McDowell County, having been recruited in Europe or in eastern American ports. Gary was segregated, but the divisions were not as sharp as in the Deep South. Black and white miners worked together and used public and company services together, but they lived apart and their children attended segregated schools. There was a great deal of interaction at times, and it was not uncommon to find southern blacks who could converse in Slavic or Polish.[34]

By 1904 a post office was established at Gary and a permanent settlement was replacing the collection of temporary shacks and huts used by the railroad workers. From 1904 to 1931 Col. Edward O'Toole supervised the development of the town of Gary. Gary soon had company offices, machine shops, commissaries, a powerhouse, and high-quality housing. The company buildings were described as stone structures with tiled roofs that were "well constructed and modern throughout as one would find in any city."[35] The houses were like those found in many cities and large towns in the area. They had electric lights, water systems, large rooms, and were placed on spacious grounds. Unlike the typical congested coal towns of the area, Gary homes were at least fifty feet apart with adequate room for gardens, lawns, and recreation. Schools and churches of many denominations dotted the hills around Gary, and a new high school was constructed in 1913. The first indoor toilets installed in an American coal mining town were installed in Gary. A sanitary committee was headed by a company physician, and residents were encouraged to participate in activities that kept the community clean and healthy.[36] The mines at Gary had an excellent safety record. Rules and regulations were strictly enforced. A 1913 article in *Coal Age* detailed the company's procedures for using basic safety measures as a guide, adding their own modifications. By the 1920s Gary was a model community among model communities. In 1923, when the U.S. Coal Commission examined the town, they gave it a score of 90 out of a possible 100 points, the highest rated town in southern West Virginia and one of the highest rated in the country.[37]

By the 1940s almost fifteen thousand people lived in Gary. The area boasted more than twenty churches, ten company stores, restaurants, independent commercial outlets, clubhouses, athletic fields, a bowling alley, tennis courts, bakeries, and theaters. U.S. Steel kept the town supplied with fresh eggs, butter, cheeses, and milk through its dairy in a neighboring county. The ethnically mixed community enjoyed social activities, sporting events, musical and cultural activities, and a variety of religious centers. Gary remained a vibrant and growing community into the 1940s. At one time there were twelve mines sites around Gary, and communities developed around each site. Each community had its own facilities and conveniences, including company stores, churches, recreation centers, and restaurants. By midcentury the town was populated by several generations of miners and their families who called Gary home. U.S. Steel was regarded as a fine company that bolstered a strong sense of community pride. The company encouraged school functions, dances, and sporting events. Different ethnic traditions thrived in Gary, centered at the many churches and social centers. The comfortable life and good working conditions enjoyed at Gary made it difficult for the UMWA to organize miners there.[38]

By the 1950s mechanization and the closing of some of the original mine sites led to a reduction in jobs. Population leveled off and then declined, but the town continued to prosper. Most of the company towns in this guidebook disappeared in the 1950s, but Gary continued to be an important part of U.S. Steel's empire for another twenty years. The town faced big changes by the end of the 1960s, however. In 1969 U.S. Steel began the process of converting Gary from a company town to an independent, incorporated city. They sold residential properties to individuals, and utilities were taken over by local companies. All services became the responsibility of the new city. In 1971 U.S. Steel oversaw the incorporation of Gary. It sold company houses and reduced services to the community. This was somewhat of a shock to the community at first, but since the mines were still open and the processing plant in operation, the town adjusted. Gary was no longer a company town, but U.S. Steel continued to play a vital role. The coal boom of the 1970s produced some new growth in Gary, and for a few years the town prospered. Once again, the expansion was short-lived. Despite a few bursts of activity in the early 1980s, U.S. Steel, and the coal industry in general, suffered. The population declined rapidly after 1982 when U.S. Steel shut down its mines and processing plant. Over the next few years they were briefly reopened, but in 1986 they closed for good. By the late 1980s, the company began demolishing company houses, mining and support facilities, churches, and community buildings. In 1991 the massive coal processing plant that once employed hundreds of people was torn down.[39] The population of Gary quickly declined and younger people had to leave the area to find work. Those remaining faced new problems. The city was unable to manage basic utilities and provide minimal services. The change affected family life as domestic disputes and alcohol-related incidents became more common. By the 1990s many of the dwellings and mining facilities had been destroyed. Today, the town is only a small fraction of what it once was.[40]

Gary lies along WV 103, south of U.S. 52 near Welch. The highway into Gary Hollow parallels the Norfolk-Southern tracks and the Tug Fork. In fact, the Tug Fork of the Big Sandy River begins its path north to the Ohio River in the mountains and streams surrounding Gary. There are a number of houses left in portions of Gary, and the current residents live scattered among the ruins. Although many of the company buildings are gone, several churches remain. In the area known as No. 3 Hollow, there is a gold-domed Orthodox church building, although the congregation has disbanded. Gary Public School, a stone collegiate Gothic–style building constructed in 1915, remains nearby. The mines have been closed and sealed and most of the mining equipment has been removed or demolished. The remains of a tram that once brought coal down from the mines is prominently located north of Gary, above WV 103. Several cemeteries throughout Gary show the great variety of ethnic groups that once populated the town. The names on tombstones bear complex eastern European names as well as the common names of white and black mountaineers who were the heart of Gary until the last years of the twentieth century.

Top: Remains of mining tram at Gary, West Virginia.

Bottom: A miner's wife and home at Gary, West Virginia.
Eastern Regional Coal Archives.

Developed in the early twentieth century, Coalwood (pop. 700), off WV 16 southwest of Welch, is another good example of a small model coal town. It was the creation of George L. Carter, who developed mining and railroad projects throughout southwest Virginia and West Virginia. Carter was one of the few native-born mountaineers to strike it rich in the coalfields, and he became one of the leading industrial developers of southern Appalachia. Carter was born and raised in Carroll County, the son of a Confederate captain, and made his first trip to southwest Virginia in 1887. He became an investor in the Pocahontas fields and organized the Virginia Pocahontas Coal Company, which opened a mine in the Coalwood area in the early years of the twentieth century. In 1905 Carter acquired the property and began building Coalwood. By 1912 the Carter Coal and Coke Company had dug a series of shaft mines and built the towns of Coalwood and Caretta. By 1915 about one thousand people lived in the immediate area. Coalwood prospered and expanded during the 1910s. Six different mine openings and about five hundred miners worked the rich "smokeless" coal seams. The town drew many workers and their families, mostly white mountaineers, and gained a reputation as a good place to raise a family. Coalwood offered high-quality housing, services, recreation, and good wages. One miner, Earl Smith, recalled that the "town had modern homes with indoor plumbing, churches, police, seven stores, a doctor, fine schools, two dentists, and even a swimming pool." Coalwood was such a pleasant place that Carter moved his family there in 1915.[41]

After building Coalwood, Carter continued investing in mining and railroad projects throughout the area. He developed the Seaboard mines in nearby Tazewell County, Virginia, and was the creator of the Virginia Iron, Coal and Coke Company (VICCO) that opened coal mines throughout Virginia and Kentucky. Carter's most important project was the building of the Clinchfield Railroad through Virginia's Dickenson, Russell, and Scott Counties. The 275-mile project eventually extended from Elkhorn City, Kentucky, to Spartanburg, South Carolina, when completed in 1915. The construction of the railroad led to the creation of Kingsport, Tennessee, developed as a model industrial city in the 1920s. Carter continued to invest in banks, mills, coal mines, and iron mills and even ran a daily newspaper in Bristol. He helped develop Middlesboro, Kentucky, and held large tracts of real estate in Bristol, Johnson City, and

Kingsport. He donated the lands that eventually became East Tennessee State University in Johnson City.

Coalwood prospered, but as the coal boom of the 1910s drew to a close Carter became more interested in other business ventures. In 1922 he sold Coalwood and nearby Caretta to the Consolidation Coal Company for $17 million. Consolidation Coal took great pride in its mining operations in southern Appalachia, and the purchase of the Carter Coal Company brought 38,000 more acres under its control. It gained ten mines in three states, most located in McDowell County. Coalwood became the headquarters of its Pocahontas–New River division. Coalwood and nearby Caretta were designated as model mining towns and received a considerable amount of attention. Clubhouses, recreation halls, and community centers were expanded. The two main shaft mines, No. 251 at Coalwood and No. 261 at Caretta, reached more than five hundred feet into the ground and together employed about one thousand miners. They were outfitted with new mining and safety equipment and electrical and telephone systems. A new ventilation system was designed to keep fresh air flowing through the deep shaft mines, and a new tipple and preparation plant were built to sort five different sizes of coal through a battery of six screens. The preparation involved chemical treatments and dust control to produce the best quality coal for market. An aerial tram carried away refuse up the mountainside and dumped it in a nearby hollow. Consolidation Coal considered the Coalwood-Caretta operations as one of their most productive and impressive communities.[42]

Coalwood maintained its reputation as a model community and continued to draw people. Under Consolidation Coal the ethnic makeup of the towns changed as European immigrants and African Americans were recruited to work in the mines. Although these mines were productive, the coal industry suffered during the late 1920s. George Carter kept his eye on coal markets during the crisis. When the Great Depression hit and Consolidation Coal went into receivership, he bought Coalwood back for a mere $4 million. The Carter Coal Company was reorganized, and George Carter's son, James Carter, moved to Coalwood to help oversee the mining operations. George Carter retired from active management in 1933.[43]

Coalwood prospered in the late 1930s and 1940s. Miners were organized in the 1930s and joined the United Mine Workers of America. When George L. Carter died, his son James W. Carter continued to supervise the mines. After 1946 the Coalwood-Caretta mines were acquired by the

Youngstown Sheet and Tube Company to supply their steel mills in Ohio and were operated through a subsidiary, the Olga Coal Company. The 1950s were the peak of production, with the mines yielding more than one million tons of coal a year and the town of Coalwood reaching a population of about 2,500 residents. During the next decade Coalwood experienced the same problems that plagued other Big Sandy region mining towns. Companies sold properties to miners and private investors, coal mining employed fewer and fewer people, and towns quickly lost population. During the 1970s Youngstown Sheet and Tube was absorbed by several corporations and the Coalwood-Caretta mines ceased operating in the late 1980s.[44] The houses and remaining properties were sold. In 1991 the Olga Coal Company laid off its few remaining workers in the area and closed its mines.[45]

Coalwood was a quiet community of retired miners until 1999, when the release of the feature film *October Sky* brought a renewed interest in the town and a small but steady trickle of tourists. The film, based on Homer Hickam Jr.'s autobiographical account, *Rocket Boys: A Memoir,* tells the story of Homer and a small group of his friends growing up in Coalwood in the late 1950s. Inspired by the launch of *Sputnik* and the Cape Canaveral team of Werner von Braun, Hickam (played by Jake Gyllenhaal) begins building and exploding rockets. Laura Dern, who portrays a high school teacher, encourages the boys and helps them with their experiments. Sympathetic neighbors and mine workers supply much-needed technical assistance and cheer the boys on. The book and film portray the complex relationship between Homer and his disappointed father, a mine superintendent who considers the entire science project a waste of time. He expects Homer to follow him in the mining industry and become a supervisor. Only an older brother, who has a football scholarship, is encouraged to pursue a college education. The boys triumph, winning medals at the regional competition of the National Science Fair in 1960. They all go on to successful careers outside of Coalwood, and Homer ends up working as an engineer for NASA, eventually working in the shuttle program.

Coalwood today looks much as it did in the 1950s, although much of the mining equipment and facilities have been removed or destroyed. In fact, visitors are often disappointed to learn that the movie was not filmed on location. The approach to the town on WV 16 is lined with company houses, many built by Carter Coal and modified by Consolidation Coal. A convenience store at a turn in the road functions as a grocery store, restaurant, gas station, museum, and tourist office. A small storage shed next to the store displays the development of Coalwood and the Rocket Boys projects. The road into town passes by the remains of the Coalwood School. (The boys went to high school in War.) Many of the homes in are two-story frame houses with end-gable roofs and front porches. Some are square, two-story homes with pyramidal roofs.

In the center of town is the former clubhouse, a large two-story white building on a stone and concrete foundation. It was originally a three-story structure with a series of dormers and windows on the top floor, but it was modified in the 1950s. The company hospital and clinic were once located alongside the clubhouse, where a small church now stands. The brick buildings across the street were the company offices and the company store. Ernest "Red' Carroll, the father of Jimmy Odell Carroll (Odell in both the book and movie) still lives in Coalwood and has taken on the role of town historian and tour guide. He can be reached through the convenience store and will gladly show visitors around the town, especially Cape Coalwood, just outside of town, where the rockets were tested.

Eleven miles down WV 16 is another coal town developed by the Carter Coal Company. Caretta is the site of a Carter Coal Company Store, built in 1912 and now listed on the National Register of Historic Places. It stands at the intersection of WV 16 and WV 12 in the center of town and is typical of company stores in the area, although several changes to the structure make it appear bigger than its original one-story design. The original brick store, with a sloping gable roof and front display windows, was placed on a stone foundation. Wooden-frame shed wings were added later, which now gives the building an uneven appearance. The brick facade visible today was added at a later date. This company store provided a great variety of general merchandise to miners and their families and also housed the company's business offices, the post office, and a doctor's office. Gasoline pumps were once located on the shaded concrete islands in front of the store. Next to the store is a white, wood frame church with four classical columns along its front entrance. A two-story brick school located across the street is still in use today. The houses in Caretta spread out in three directions from the center of town, and there were once two smaller company stores to serve residents.[46]

The history of Caretta parallels that of Coalwood, but Caretta became the industrial center of the area and one of the largest operating mines in southern West Virginia. South of town on WV 16 is the Olga Coal

Clubhouse, church, and houses in Coalwood, West Virginia.

Company Caretta Mine Complex, a historic complex with steel-frame mining structures and brick support buildings. There are remains of a supply building, maintenance shop, hoist house, and offices.[47]

Bartley

The coal town of Bartley (pop. 350) is located on WV 83, in the southern part of McDowell County, about eight miles from War. WV 83 can be reached by continuing down WV 16 from Coalwood or by following WV 80 south from Iaeger. Bartley, which lies on the banks of the Dry Fork River, south of the Tug Valley, is a typical coal town but has the distinction of being the site of the worst coal mining disaster in the greater Big Sandy Valley. In fact, the explosion at Bartley more than sixty years ago was one of the worst in the history of West Virginia.

Bartley was settled in the early 1900s and was named for the creek that runs alongside the town. By 1909 the Bartley Coal Company was operating a series of small drift mines in the vicinity. Coal mining drew people to the area and the company continued operations until the late 1910s, when it chose to lease the land for mining. Bartley began to grow after 1922 when the Pond Creek Pocahontas Coal Company opened several new shaft mines and hired many new miners. The No. 1 mine was in operation by 1923 and nearly three hundred men were employed at the site. The No. 1 was a shaft mine where workers were taken down 584 feet to work the Pocahontas No. 4 coal seam. No. 1 proved to be a very productive mine and put out about fifteen hundred tons of coal a day. Two other mines were later opened at Bartley, the No. 2 in 1930 and the No. 3 in 1932.[48]

As the Pond Creek Pocahontas Coal Company expanded, Bartley drew more residents and developed into a thriving community. In 1926 an elementary and middle school were opened. Churches were built and company stores served the miners and their families. The company provided good housing and services, but Bartley was not a model community. By the late 1930s Bartley was a typical company coal mining town, with a population of about one thousand. The town was divided into three areas or camps, centered around the three mines in operation. It was primarily a residential town but also had a company store, post office, medical clinic, dentist's office, and company offices.[49]

In the winter of 1939 Bartley was still recovering from a tragic accident that had taken place earlier in the year. On the way to Big Creek

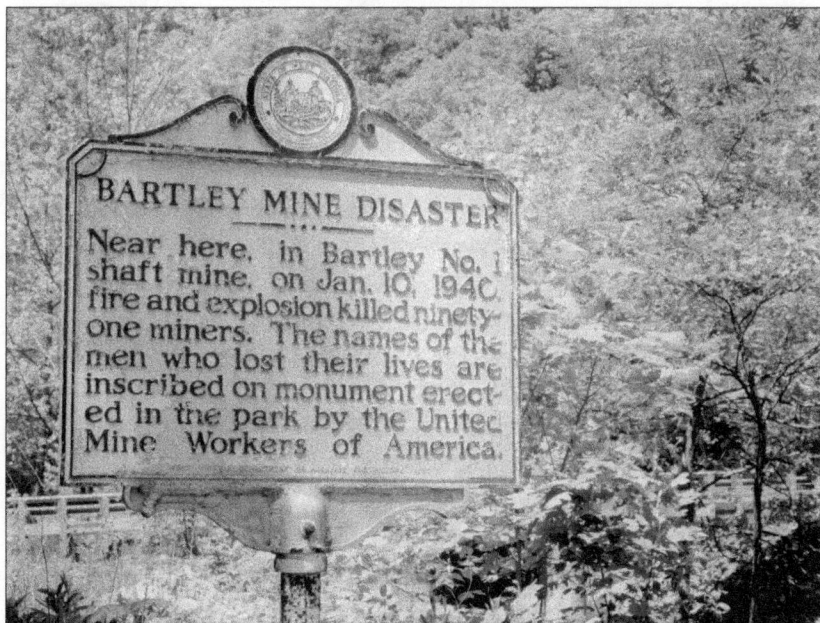

A West Virginia historical highway marker indicates the site of the Bartley Mine Explosion.

A United Mine Workers of America monument dedicated to the ninety-one miners who lost their lives in the Bartley Mine Explosion.

High School in War, a school bus driver lost control of the vehicle after one of its steering components broke. The bus plunged over an eighty-three-foot embankment and crashed. The wreck left six students dead and sixty-six others injured. Although the town was in mourning, the worst was yet to come. On 10 January 1940 there was an explosion deep in the No. 1 mine at about 2:30 P.M. People outside the mine immediately knew something had happened when the ground rumbled and clouds of dust flew from the mine shaft.

Seven company rescue teams entered the mine and encountered scattered bodies, many burned beyond recognition. As they advanced into the mine, rock, twisted steel, and coal blocked their path. It was the beginning of a long, slow process. In the meantime, hysterical family members gathered around the mine site desperate for information. Guards were placed near the mine, onlookers were forced back, and the area was sealed off. To make matters worse, a series of small explosions followed as pockets of gas built up. The crowd, which had now grown to almost two thousand, began to panic, thinking the rescue crew was now in danger. As the crew members made their way through the mine, they found more bodies.

Although it was clear there were no more survivors of the blast, the process of locating and identifying the bodies took almost two weeks. Wrapped and doused with formaldehyde to slow their decay, the bodies were gradually brought to the surface and moved to local funeral homes. Along with the bodies, rescue workers found quickly scrawled notes miners had written in the last minutes of their lives. Some stated what type of funeral they desired; others simply expressed their love for their wives and families.[50]

The explosion at Bartley took ninety-one lives. It directly impacted seventy-five families and left 278 children fatherless. In a small mining town such as Bartley, the real impact could never really be measured. Everyone knew the men who had perished in the explosion. They knew their wives and children and shared in their suffering. Most of the men involved in the rescue work lived and worked in Bartley, and the bodies they removed from the mines were those of their friends, coworkers, and family.

After the tragedy, many of the survivors moved away. Widows and children took refuge in neighboring towns, and some left the coal mining region, never to return. Mining continued at Bartley after the tragedy, but the depressed coal industry of the 1950s took its toll. By 1958, the mines were worked sporadically and few men were regularly employed.

Today, Bartley is home to several hundred people and has gained a quiet, somber reputation. Many of the few remaining residents are retired miners, relatives, or acquaintances of those who lost their lives. There are about one hundred houses scattered along the creek in Bartley between the mine sites and the center of town. A West Virginia historical highway marker near the site of the mine, off Bartley Hollow Road, tells the story of the explosion. Just south of the marker, in front of the Bartley Church of God, a monument erected by the United Mine Workers of America lists the names of the ninety-one men who lost their lives in the tragic explosion.[51]

To continue on the Coal Heritage Trail, return to U.S. 52 and head east from Welch. Several small mining towns, Landgraff, Algoma, and Powhatan are along the route. Eckman is the former site of Norfolk & Western's Eckman Yard, where a large roundhouse once serviced mallet steam locomotives.[52]

On U.S. 52 about twelve miles southeast of Welch, Elkhorn (pop. 250) is an example of a quality coal town built in the late nineteenth century and is one of the most intact mining towns in McDowell County. Although most of the large company buildings and mining equipment are gone, it still has many residences and a few other interesting structures. Elkhorn was once a very diverse community, dominated by immigrant workers and their families. It was created by a benevolent coal operator who developed Elkhorn as a small model mining community.

John J. Lincoln played a fundamental role in the building, development, and operation of the town of Elkhorn and its mines. A Pennsylvania Quaker, Lincoln was born in 1865 in Lancaster County and graduated from Lehigh University in 1889. He was a survey engineer for the federal government until 1892 when the Crozer Land Association offered him an opportunity to move south to the Pocahontas coalfields and help develop Elkhorn Valley mining operations. He was soon chief engineer, and during the next few years he became vice president and later general manager of the Crozer Land Association and the Page Coal and Coke Company. Lincoln was influential far beyond the mining town of Elkhorn. He lived in the area for almost sixty years and held many positions in other companies, in the Republican Party, and in state and local government.[53]

When Lincoln arrived in West Virginia, the railroad had just opened the coalfields a few years earlier and conditions were primitive. There were problems with gambling, prostitution, and alcohol, and Lincoln stated in a 1938 memoir that "seldom a weekend passed without several shootings and sudden deaths." When he visited the nearby mining town of Pocahontas, Virginia, he noted that "about every other building was a saloon."[54] He was shocked by the frontierlike atmosphere along the railroad lines and became determined to create better living conditions for workers. As he oversaw the development of the Elkhorn area and its mines, Lincoln developed the town as a model community and strove to provide an oasis of civility for his employees and their families. The town had tree-shaded streets, spacious grounds, and public areas. Houses were well equipped and neatly painted green and white.

Three coal companies leased land from the Crozer Land Association and operated mines in Elkhorn: Crozer Coal and Coke, Houston Coal and Coke, and Upland Coal and Coke. The mines at Crozer Coal and

Coke were quite advanced for their day. In the 1890s they used Jeffrey cutting and drilling machines. The companies mined the rich Pocahontas No. 3 seam, which ran along Elkhorn Creek. The mines were so productive that by the 1930s they had produced more than sixty-five million tons of high-quality coal.

The town of Elkhorn grew and soon had three company stores, a freight and passenger station, hotel, restaurant, barbershops, and a meat warehouse. In addition, each coal company ran its own powerhouse. Italian stoneworkers made coke ovens, and Hungarians, Slavs, and Russians arrived to work in the mines. By the 1910s Elkhorn was a town of more than one thousand people, a rich mixture of eastern Europeans, African Americans, and white mountaineers. Lincoln took a keen interest in education and provided high-quality schools for his workers' children. At one time he served as the president of Elkhorn District's Board of Education. Along with formal education, there were also recreational and social activities in Elkhorn. The Elkhorn Orchestra performed annually on the lawn of Lincoln's estate. During Lincoln's time, Elkhorn remained a fine place to live and work. The mines were still in operation at the time of his death in 1948. Afterward, the company continued mining coal at Elkhorn but slowed production during the 1950s. The Crozer operations were purchased by the Consolidation Coal Company in 1963 and worked until the 1970s.[55]

Elkhorn lies alongside Elkhorn Creek, off U.S. 52, and a visit today reveals much of its coal mining past. The town has several buildings still in use and retains a small population, many of whom are descendants of the men who worked the mines earlier in the century. In fact, the town has periodic reunions. The road from the highway leads directly to what was once the center of town. An old Norfolk & Western 2850 caboose stands in a small park alongside the railroad grade where the depot was once located. Across from the park is the original 1888 post office building, in use for more than one hundred years and recently nominated to the National Register of Historic Places. It appears much the same today as it did when it opened. Unlike most area post offices that were located in company stores or offices, Elkhorn's is in a small separate building. This sixteen-by-thirty-two-foot wooden frame building stands barely eighteen feet high with its peaked roof. This post office saw countless thousands of letters pass through it, some from nearby towns and communities, others in Cyrillic lettering bound for faraway eastern European hometowns.[56]

The post office at Elkhorn, West Virginia, has been operating for more than 110 years.

John J. Lincoln, the developer of Elkhorn, West Virginia. Eastern Regional Coal Archives.

The land alongside the post office was once the heart of Elkhorn. Here stood a school, restaurants, a barbershop, apartments, foreman's quarters, and the Swift and Company meat warehouse and store. To the west was the two-story brick and stone Elkhorn Grade School. On the other side of the post office is John J. Lincoln's home, listed on the National Register of Historic Places. The Lincoln house was built in 1899 on a spacious two-and-one-half-acre yard at the western end of town, surrounded by poplar and maple trees. It is a Victorian style home with a rough stone foundation and shingle and copper portions in the roof. Clapboard and wooden shingles cover the exterior, and a porch wraps around the front of the home. In the rear is a glass-enclosed porch. Much of the home's interior remains intact, and some of the original furnishings are still in use. There is a large craftsman-style stone fireplace and decorative mantels. Beyond the house is a small single-story clapboard building that once served as a school for Lincoln's children.[57]

Following the railroad grade beyond the school, there are the remains of a machine shop, a water tank, and several small buildings along the tracks. Houses lie along the main road as it winds through Elkhorn and climbs the surrounding hills. Most are two-story wood frame with attached front porches facing the narrow sidewalks and railroad tracks. Many have a downstairs addition at the rear. Several residential neighborhoods with churches are located on the hillsides of Elkhorn. One of the interesting structures is St. Mary's Russian Orthodox Church, built for the Ukrainian immigrant community in 1911. The Byzantine white wooden church sits prominently on the hillside at the east end of Elkhorn, overlooking the community below. It has a stone and brick foundation, a series of steps leading up to the front entrance, and a gold onion dome that is visible throughout the town.

Beyond McDowell County

For travelers who wish to continue exploring the southern Appalachian coalfields beyond the Big Sandy region, Mercer County (pop. 62,980) and nearby Tazewell County (pop. 44,598), Virginia, offer fascinating coal mining sites. The Flat Top–Pocahontas coalfields that straddle the West Virginia–Virginia border had some of the richest coal seams in the region. A great coal mining boom followed in McDowell and Mercer Counties and across the state line in Pocahontas. These fields contained some of the best bituminous coal available, in seams up to thirteen feet deep. In 1884 the Flat Top Land Company set up a company headquarters and laid out the new town of Bramwell. By 1891 there twenty mines and more than 900 coke ovens operating in McDowell County. Mercer County had thirteen mines and 710 coke ovens. More than 3,500 miners were employed in these new mine sites.[58]

Bramwell

Mercer County's Bramwell (pop. 426) is located on WV 120, off U.S. 52. It was once known as a village of millionaires where the men who developed area coalfields lived in great splendor. Joseph H. Bramwell, a civil engineer from New York who founded the town in the early 1880s, became the first postmaster and named the town for himself. He invested heavily in the mining industry and became the first president of the Bank of Bramwell. Four large coal and coke companies dominated the area around Bramwell: the Mill Creek, the Buckeye, Booth Bowen, and Caswell. The Bank of Bramwell became the financial center for mining operations and land enterprises throughout southern West Virginia. Locals told stories of wheelbarrow loads of cash being transferred from the bank to the railroad depot. Many fortunes were made and lost in the boom/bust cycles of the coal economy, but unlike many of the new millionaires, Bramwell survived. In the 1930s he cashed in his holdings and moved to Switzerland.

Early Bramwell grew fast and soon absorbed the nearby mining camps of Coopers and Freeman. The area around the town grew as small mining settlements drew thousands of workers. Shops in the downtown commercial district prospered, and four thousand people lived in the hillside communities surrounding the town. The Norfolk & Western Railway brought mining equipment as well as imported exotic goods to the local depot. As a hub of the mining boom along the Bluestone River, it soon became a residential oasis for wealthy coal barons who built beautiful, spacious homes as profits soared.

Bramwell is a well-preserved town and retains much of its elegant turn-of-the-twentieth-century charm. Although many of the regional commercial centers described in this guidebook have mansions and elegant residences, none has the atmosphere of Bramwell. The quiet residential streets lined with spacious, luxurious mansions are unparalleled in the region. In 1983 Bramwell was added to the National Register of Historic Places.

The town hall and tourist information center, a two-story frame building constructed in 1889, is located at the entrance to town on WV 120 and is one of the oldest surviving buildings in Bramwell. Across from the town hall is the reconstructed Norfolk & Western depot, which is scheduled to be the new tourist information center and the future hub of the Coal Heritage Trail. Beyond the depot, WV 120 becomes Main Street and enters the commercial district of the town. Hotels, offices, saloons, and shops once lined both sides of the street until a devastating fire wiped out the commercial center of Bramwell in 1910. On the south side of the street only the elegant Bluestone Inn remained. It was later demolished in the 1930s. Following the fire, the north side of Main Street was rebuilt with brick, stone, and mortar. Today, six buildings, all two- or three-story commercial structures, line the street. Across Bloch Street is a private residence that once housed the offices of the Pocahontas Coal and Coke Company. It is a two-story building with a small gabled projection in the center of the roof, facing the street. This was the site of the company office that designed the town and developed mining in the area. Inside it has elaborate mahogany woodwork and three vaults that have been converted into baths and storage.[59]

The Bank of Bramwell was located across the street in a rectangular two-story gabled building with an arched doorway. Incorporated in 1889 and constructed in 1893, the bank was said to be one of the richest in the country for its size. It maintained a plain exterior and never posted a sign out front. The interior, however, was quite different. Red oak paneling and ornate wood carvings give the inside a lavish, elegant appearance. White Italian and colored marble tiles were installed in 1901. The bank's first cashier and most influential president, Isaac T. Mann had his office next to the board of directors' room and his name is still brass-lettered on the door. The Bank of Bramwell was an extremely active bank considering its size and location. During the First World War, it sold more bonds than any other bank in the nation. It also financed large projects outside of Bramwell, including the Burning Tree Country Club in Washington, D.C.

The Bramwell Presbyterian Church was built in 1903 and was donated to the town by Isaac T. Mann. It is an English Gothic stone structure with a steeply pitched roof and a square tower that faces Bloch Street, dominating the front entrance to the church. The building is made of local bluestone, a sandstone mixed with coal dust, and is said to be modeled after a small Welsh cathedral. Italian stone masons who came to the area to work in the mines cut and laid the stone.

Bramwell is best known for its beautiful mansions and houses. A walking and driving tour brochure details twenty-one historic sites, including twelve late Victorian–era homes. The most prominent is perched on a hillside, overlooking the Bramwell Presbyterian Church and facing the commercial district. It was built by Philip Goodwill, a Pennsylvania-born coal operator who developed mines in northwestern Mercer County and established the Goodwill Coal and Coke Company, as well as the town of Goodwill. Goodwill left Mercer County and became a lawyer but never practiced. He returned as a general manager of the Goodwill mines in 1887, later selling his properties and becoming president of Bramwell's Pocahontas Company in 1905.

Goodwill's house was originally built in 1894 as a simple two-story frame dwelling. As his wealth increased he decided to make major changes by adding a third floor, a semicircular turret, and a long encircling front porch. On the third floor were guestrooms, a game room, and a spacious ballroom where the Goodwill's entertained Bramwell's elite. Goodwill's children played in a small tower room in the top of the turret. Much of the foyer is paneled in beautifully rich oak, and the dining area has thick oak beams running along the high ceiling. Like many of Bramwell's residents, the Goodwills enjoyed an elegant and comfortable life as long as the coal business boomed. When the Pocahontas Company dissolved in 1909, hard times came to the family. Phillip died in 1916 and his family struggled to survive on investments and assets. His wife, Phoebe, lived in the house another eighteen years before moving to the Bluefield's West Virginia Hotel, where she stayed until her death in 1953. None of the children continued in the coal business.

Beyond the Goodwill house, off Duhring Street, was the home of Pocahontas coal operator William Thomas. The Thomas family wanted a house that would reflect their British heritage, so they chose the English Tudor revival style for this three-floor eighteen-room home. It was said that the timber and interior wood paneling were imported from England when it was built in 1909. Mrs. Thomas hired masons from Italy to lay the stonework for the house and its surrounding walls. The house has several chimneys, prominent gables, a front porch, and a red tile roof. The project cost more than $95,000 to complete. The interior features a large stairway that splits into two separate staircases that lead to a landing under a huge stained glass window. William Thomas died in 1918, but his widow lived in the home until her death in the 1940s.

Bramwell's best-known early resident was Isaac T. Mann, one of the most powerful financiers in the West Virginia coal business. He bought a large house on South River Street in 1900 and had it remodeled to suit his tastes. The long and narrow three-story structure has a turret and a rounded porch with tall, slender columns. The third floor was used for the children and a rear section of the building functioned as the servants' quarters. Mann was president of the Pocahontas Fuel Company for more than thirty-five years and amassed a fortune of more than $80 million by 1930. Unfortunately, he lost most of it, along with his health, during the Great Depression. Other interesting homes include the Cooper house, a lovely Queen Anne with a copper roof; the Hewitt house, a prairie school–style home that is now a bed and breakfast; and the Hickman house, which features an Italianate design.

The Oak Hill Cemetery, originally affiliated with the Bramwell Baptist Church, is a beautifully landscaped cemetery surrounded by a stone wall and ornamental wrought iron gates. Here lies many of Bramwell's founding residents, including Isaac T. Mann, John Freeman, and John D. Hewitt. An iron gateway leads to the separate section where African Americans are buried.

Although Bramwell is best known as home of the rich and famous, most of the area's residents were common workers.[60] South of Bramwell lies the town of Cooper, where there are several clusters of small company houses remaining. Most have been modified over the years but are generally one-story homes with front and rear porches. North of Bramwell is Freeman, where workers' and miners' houses line Simmons Avenue. Southern West Virginia's mining industry drew a large number of African American workers, and there were segregated neighborhoods where hundreds of black miners and their families lived. West of Simmons Avenue was the African American section of town and the Bluestone Baptist Church. In the center of Freeman is the former Booth-Bowen Company store, a two-story end-gable building with a large first-floor storefront. Several of these communities were added to the National Register of Historic Places in 1995.

Several of Bramwell's African Americans became community leaders and helped shape life in Bramwell and beyond. Dr. William Alexander

Right: Isaac T. Mann, one of Bramwell's most prominent early residents. Eastern Regional Coal Archives.

Below: Philip Goodwill's home, Bramwell, West Virginia.

Holley, a native of Wytheville, Virginia, and a graduate of Howard University School of Medicine, came to Bramwell in 1892. He helped organized social and educational functions in the area and founded West Virginia's first colored Masonic Lodge, Lodge No. 1, in Bramwell. He went on to organize the Statewide Grand Lodge and later became the first state Grand Master. The Golden Rule Association, one of the most active black fraternal organizations in the early twentieth century, had its headquarters in Bramwell. The association thrived under the leadership of Rev. R. H. McKoy, serving blacks throughout West Virginia, Kentucky, and southwest Virginia. It had more than 5,280 members and fifty-four subordinate lodges and paid out more than $13,000 in death and sick claims.[61] Poetess Ann Spencer lived in the black area of Cooper in the 1890s. She went on to attend the Virginia Seminary and Normal School at Lynchburg, graduating valedictorian in 1899. Spencer became a librarian and established an African American library. She became well known for her writing and made regular contributions to *American Negro Poetry*. She and her husband, Edward, also helped organize civil rights activities. The Spencer home was a popular stop for blacks passing through Lynchburg during the era of Jim Crow. Some of their guests included W. E. B. DuBois, Langston Hughes, and George Washington Carver.[62]

At one point, Bramwell claimed to be the richest town of its size in the country, with fourteen resident millionaires. It enjoyed great prosperity until the 1930s when the bank, coal companies, and many of its residents became victims of the Great Depression. Today, Bramwell has many interesting homes from the turn of the twentieth century that can be seen on a leisurely walk through the town.

Some of the first mines in the area were located near Bramwell. A West Virginia historical highway marker, located at the junction of WV 20 and WV 120, indicates the site of the John Cooper's mine, the first to ship coal from the Pocahontas coalfields over the Norfolk & Western Railroad in 1884.[63] West of Bramwell, on WV 120 near the Virginia border, a West Virginia historical highway marker indicates the spot where blacksmith Jordan Nelson mined and made some of the first commercial sales of coal from the region. He was selling coal for one cent a bushel in the 1870s when Norfolk & Western surveyors visited the areas. Within ten years the Pocahontas Mine was opened and coal was taken out along the Norfolk & Western tracks.[64]

Pocahontas

The first major coal mining town in the area was Pocahontas (pop. 441), established in 1881. Today it remains a well-preserved coal town with many historic buildings and an exhibition mine. Although the coal deposits along today's Virginia–West Virginia border were first noted by Dr. Thomas Walker during his explorations of the area in 1750, it was more than 130 years before the area was opened to mining. In the 1870s former Confederate major Jedidiah Hotchkiss promoted the mineral wealth along the Bluestone and Big Sandy Rivers and hired Isaiah A. Welch to do further survey work in the area. Welch reported great coal seams, some as thick as thirteen feet, and visited a few sites where local pioneers were mining coal. Following the Philadelphia Centennial Exposition of 1876, where Hotchkiss displayed samples of the coal, he convinced a group of Pennsylvania investors to found the Southwest Virginia Improvement Company and begin developing the Bluestone River area for mining. In the early 1880s a new settlement called Pocahontas (after the legendary Virginia Indian Princess) was built in the wilderness, awaiting the arrival of the railroad.

An early camp was set up to house the workers preparing the new mine site. Large numbers of Hungarian, Swedish, and German workers were recruited in New York, and shanties and tents appeared at the opening of the No. 1 Pocahontas mine. Offices and houses appeared shortly thereafter, even though the railroad was still more than fifty miles away. It took time for the new line of the Norfolk & Western Railroad to reach Pocahontas, but the first train finally arrived in March 1883. By then, more than two hundred coke ovens were operating, houses were scattered around the mine site, and stores, saloons, and company offices were in place. Hundreds of workers flocked to the area, and small mines opened in surrounding settlements. Saloons, gambling, prostitutes, dance halls, and occasional shootouts soon gave Pocahontas, one of the only wet towns in dry Tazewell County, a reputation as a wild frontier town.

Small three- and four-room houses were constructed for workers around the original mine. Along Railroad Avenue and Powell's Alley, two-story four-pen dwellings were built for company managers and supervisors. The most elaborate residences were on St. Clair Street and featured fireplaces and running water. More than one hundred workers' houses were clustered around May Street. Centre Street became the town's commercial center, with company offices, hotels, a post office, company store, and numerous small businesses. There were more than seventy-five commercial establishments in Pocahontas by 1895.[65]

The Pocahontas mines were extremely productive but were also the site of tragedy. On 13 March 1884 a major mine explosion rocked the town. A rush of hot air, coal dust, and fragments blew out of the mine portal, destroying houses and trees, tipping coal and railroad cars, and damaging much of the equipment around the fan entry. The explosion killed 114 miners, foremen, and workers in mine No. 1 and caused extensive damage around the mine site. A second explosion on 3 October 1906 killed thirty-six miners.

Despite these tragedies, Pocahontas prospered and grew. By the 1920s, more than thirty-five hundred people lived in the community and most construction had been completed. But within ten years, Pocahontas entered a long period of decline. The coke ovens were phased out by the 1920s, and during the 1930s mine waste was dumped at the site, burying the remains of hundreds of coke ovens. The peak year of coal production was 1927, when the three mines operated in Pocahontas, but the great mining boom slowed down by the time of the Great Depression. Although mine No. 1 closed, two others continued operating until 1955. During the seventy-three-year period that the mines were open, they produced more than forty-four million tons of the world's highest quality coal.

Pocahontas is a well-preserved town, although many of its buildings have suffered in recent years. It has an interesting assortment of houses, public and commercial buildings, and religious sites. The main residential areas are located in the center of town and consist of one- and two-story company-built frame houses. Most are two-story duplexes with shed roof porches. Along Water Street, east of Centre Street, are a series of larger homes built for managers and company supervisors. Some are Queen Anne style with cross gables and irregular floor plans. There are a few remaining small coal houses in front of some residences. An early-twentieth-century two-story brick schoolhouse, in poor condition, overlooks the town.[66]

The Opera House and Old Town Hall building is located on St. Clair Street in the center of town. This eclectic two-story brick structure was built in 1895 and served as the first theater in the area. It featured a variety of entertainers and performers during its heyday, including the Midnight Follies and the Budapest Hungarian Opera. After many years of neglect, it was renovated and reopened in 1973 and today serves as a site for many civic and social events. The commercial district of Pocahontas once spread along Centre and Water Streets and still has a variety of late-nineteenth-century two-story brick buildings. Early Pocahontas was known throughout the area for its wild, frontierlike atmosphere and wide selection of saloons and brothels. Three-story houses, with saloons downstairs and brothels on the upper floors, were common around the town. Fancy establishments that catered to the elite were located near the depot; other more primitive places were found on the surrounding hillsides. The Silver Dollar Saloon on Centre Street, one of twenty-seven saloons in Pocahontas, was opened in 1894. This two-story brick building, next to the site of the Elkhorn Saloon, featured a unique concrete and marble floor with real silver dollars. The original company store, a two-story frame building with a mansard roof on Water Street, was built in 1883.

There are several churches in Pocahontas. Most are wood frame buildings, but the Baptist Church at the west end of Water Street is of native-stone construction. On the hillside overlooking the town is St. Elizabeth's Catholic Church, known for its beautiful restored frescos. The wood frame church was built in 1896 under the guidance of a French priest, Father Emil Olivier, and named for Saint Elizabeth, the patron saint of Hungary. In 1919, a Cincinnati artist, Theodore Brasch, painted

The once-thriving commercial district of Pocahontas, Virginia.

a series of ten eight-by-twelve-foot life-sized murals depicting biblical scenes that still grace the interior of the church today.[67] The church became an independent parish in 1987 and is still in use today. The Jewish Synagogue building is a beautiful two-story brick structure on Church Street. The Pocahontas Hebrew Congregation was organized in 1892, and by the early years of the twentieth century there were as many as fifty Jewish families living in town. Shortly after 1900, the synagogue was constructed and regular services were conducted on Friday evenings. Many Jews ran the saloons and stores in town and prospered from the booming coal economy. After 1916, when Pocahontas prohibited liquor sales, area businesses closed and Jewish families left the town for Bluefield. By the 1920s there were few Jews left in Pocahontas. In 1951 the building was sold to raise money for a new synagogue in Bluefield.[68]

The main attraction today is the Pocahontas Exhibition Mine, listed on the National Register of Historic Places. It was designated as mine No. 1, being the first commercial coal mine opened in the Pocahontas fields, and operated until the 1930s. The large one-story stonewalled powerhouse is located next to the mine entrance and serves as a mining museum and visitors' center. The present entrance was once the fan house, where fresh air was blow into the mine. The fan house, a brick and rough stone structure covered with an asphalt roof, is topped by a brick archway and a

The Pocahontas Exhibition Mine.

sign identifying the mine. Next to the entrance is an outcropping of the incredible thirteen-feet-thick Pocahontas No. 3 coal seam. Until 1970 visitors could drive their automobiles through the mine. Today, visitors are taken into the mine through the fan house on a walking tour and exit through the main mine portal. The forty-five-minute escorted tour includes an explanation of the geology of the region, the history of coal mining, an overview of mining techniques, and the story of the tragic 1884 explosion. More than one million visitors have passed through the mine since it was first opened to the public in 1938. Tours are offered seasonally and there is an admission charge.[69]

Bluefield

Southeast of Bramwell on U.S. 52 is Bluefield, West Virginia, and its sister city, Bluefield, Virginia. Bluefield, West Virginia (pop. 11,410), is the site of one of the country's largest railroad classification yards and a major transfer point for coal heading east to the coast or north into the Ohio Valley. It was established in 1881 with the arrival of the Norfolk & Western Railroad but took a few years to grow as a regional commercial and transportation center. When a post office was chartered in 1887, the town was home to fewer than one hundred residents. It became a stop on the Norfolk & Western line and soon had a depot, a roundhouse, machine shops, and side tracks laid out for its yard. As the Flat Top–Pocahontas coalfields expanded, Bluefield grew rapidly, with five thousand residents by the turn of the century. Norfolk & Western trains hauled coal hoppers into the rail yard day and night, and the powerful locomotives were serviced in an immense engine roundhouse.[70] Coal town company stores purchased goods from the massive warehouses that occupied the west end of the yard. The railroad employed thousands of people in the Bluefield area, fueling the commercial growth of the city. Above the rail yard a crowded brick downtown commercial center developed, housing countless stores, shops, warehouses, banks, and coal company offices. By the 1920s the central business district featured large ornate concrete and steel masonry buildings. Along "The Avenue" were trendy shops, restaurants, and theaters. Entertainment and shopping attracted miners and their families from the many area coal towns, making weekends a fabulous display of people of many ages, races, and ethnic backgrounds.[71]

More than twenty-five thousand people lived in Bluefield by the 1940s, including a thriving African American community. Bluefield State College, established by the West Virginia State Legislature as the Bluefield Colored Institute in 1895, prepared African American teachers for work in the state's public schools. The college expanded its programs in the early twentieth century and became one of the few educational institutions in the area to serve the black community. It was renamed Bluefield State Teachers College in 1929 and Bluefield State College in 1943. The college is off U.S. 52, and two West Virginia historical highway markers are located on the grounds.

Bluefield's downtown commercial district has a variety of early-twentieth-century buildings. The *Daily Telegraph* building stands at 412 Bland Street. This neoclassical brick building, with its prominent terra-cotta Corinthian columns and keystone arch entrance, was built in 1916 and has been the home of southern West Virginia's only daily newspaper for more than seventy years. The Bailey Building, a six-story brick classical revival office building located at 704 Bland Street, was constructed in 1923 and was the site of the notorious Baldwin-Felts Detective Agency. The West Virginia Hotel, located at Federal Street and Scott Street, was a twelve-story luxury hotel built in the 1920s. It towers over the railroad yard and remains the tallest building in southern West Virginia.

Despite a declining population in the late twentieth century, Bluefield continues to serve as a major railroad center for the Flat Top–Pocahontas fields. Four and one-half miles of the length of central Bluefield are split by the more than twenty tracks of the Norfolk-Southern yard, a gravity switching yard. A stone wall and Princeton Street separate the yard from the downtown commercial district of the city. Rail fans consider this to be the quintessential mountain railroading town. The yard runs through the center of the city and there are many places to view the countless coal hoppers hauled and switched by powerful Norfolk-Southern engines. At the Mercer County Fairgrounds outside of Bluefield there is a Norfolk & Western 2-8-0 Steam Locomotive on display.[72]

Bluefield is also the site of the Eastern Regional Coal Archives, housed in the Craft Memorial Library. It has an extensive collection of documents, photographs, and coal company records that deal with the southern West Virginia coalfields. The archives originated in the accumulation of materials for the 1983 Pocahontas Coalfield Centennial Celebration. The materials were later transferred to the Craft Memorial Library, and the archives have since been supported by the A. R. Mathews Foundation and the public library system. There are displays of photographs in the library, and the archives section, located upstairs, has a vast collection of materials for those who wish to research the coal mining heritage of southern West Virginia.

Top: The trendy "Avenue" in Bluefield, West Virginia, circa 1930. Eastern Regional Coal Archives. Bottom: A massive coal marshaling yard cuts through the center of Bluefield, West Virginia. Eastern Regional Coal Archives.

5   Coal Towns in East Kentucky

Up the Big Sandy River into Kentucky

The Big Sandy River drains into the Ohio River in Boyd County, Kentucky. From here, travelers can head south into the east Kentucky coal country by following U.S. 23 or taking Interstate 64 east and continuing on U.S. 52 into West Virginia. The northern end of east Kentucky has little coal but has played an extremely important role in the development of the iron industry, water and rail transportation, and mining.

The Hanging Rock Iron Region

In the 1820s the Hanging Rock Iron Region, which encompasses parts of northeastern Kentucky, southern Ohio, and West Virginia, emerged as a major center of the nation's iron furnace industry. Hanging Rock iron furnaces were usually thirty to forty-five feet high, made of cut stone, and fueled with locally produced charcoal. They were located near streams to use water power for operating the bellows. The process involved mixing iron ore and limestone with charcoal, then burning the combination in a stone stack to produce a workable iron. Large forge hammers made wrought iron for blacksmiths, who then turned the metal into tools and common implements. Pots and stoves were cast in sand molds in houses near the furnaces. Iron production soared by the 1830s as more construction supplies, architectural decorations, consumer goods, and tools were made of iron. The biggest demand came from the new railroad industry

as tracks were laid throughout the Northeast. By the 1840s iron producers had begun to switch from charcoal to coal as a fuel.

By then the Hanging Rock Iron Region was the leading iron producing region in the United States, but this distinction was short lived as
much larger, efficient furnaces were being built in urban areas. These furnaces, developed by companies rather than individuals, made use of new
technology, cheap Pennsylvania coal, and lower transportation costs to
produce a better quality iron. Beginning in the Pittsburgh area, company-
owned furnaces spread into Kentucky, and by the 1850s John Means's
ironworks in Ashland and Daniel Hillman's operations in the Cumberland area began to control the market. Although the conversion to coal
allowed many of the Hanging Rock iron furnaces to survive, the Ashland
ironworks soon dominated.[1]

The Hanging Rock Iron Region has several interesting sites related to
the development of the iron industry and mining. The most impressive

Remains of the
iron furnace at
Mt. Savage, Kentucky.

iron furnace ruin in the area is located in Carter County at Mount Savage (pop. 150). Mount Savage is located about one and one-half miles east of Hitchins, on KY 773. The original town of Mount Savage no longer exists, but the furnace ruins are large and imposing, located right alongside the highway near a small cluster of mobile homes. The town was originally situated along Strait Creek, a tributary of the Little Sandy River. Robinson Biggs hired a Prussian stonemason, John Fauson, to build an iron furnace in 1848 on the lands of Edward Savage, and the small community of Mount Savage soon grew around it. The iron was loaded on wagons and hauled by teams of oxen to the shores of the Ohio River, where it was taken north. The furnace produced an average of fifteen tons of pig iron daily that could be used for tools, plows, rails, and machinery. By the 1880s local limestone, charcoal, and iron ore were growing scarce and production declined. The furnace operated for thirty-seven years, firing its last blast in 1885. Mount Savage thrived for a few more years, but most of the town's residents moved away in the early twentieth century.[2] The Mount Savage Furnace, noted by a Kentucky highway marker, is the best preserved and one of the most beautifully made native stone furnaces in the entire region.[3]

Rush

Located along Rush Creek about ten miles southwest of Catlettsburg on KY 854, Rush (pop. 200) is one of the few nineteenth-century coal towns remaining in the region. It can be reached by following U.S. 60 south to KY 854. This small agricultural town was settled in the early nineteenth century as Geigerville but grew in the 1870s with the mining of a nearby coal seam, becoming one of the first coal towns in east Kentucky. It prospered until the early 1890s, when its coal reserves ran out. Little else is known about the town, and today Rush is a small community with little indication of its coal mining past. There are a few small one- and two-story frame houses remaining from the late nineteenth century that have undergone extensive renovation.[4]

To begin exploring the Big Sandy River Valley, follow U.S. 23 south from Boyd County. Boyd County (pop. 48,500) was created in 1860 from parts of Greenup, Carter, and Lawrence Counties and was named for Lin Boyd (1800–1859) of Tennessee, who came to western Kentucky in his youth, entered politics, and served as a U.S. congressman. Boyd County has had limited coal mining, but it was central in the early use of coal and the transfer of coal to northern markets along the Ohio River. Russell, Ashland, and Catlettsburg have served as major rail and river transfer points for timber and coal from southern counties. The early iron industries of the region led to a thriving steel industry by the twentieth century.[5]

Russell

The town of Russell (pop. 3,645), between the Ohio River and U.S. 23, grew around the Russell Terminal, the largest yard run by a single railroad in the world and the main transfer point for coal headed north to the industries and businesses of the Great Lakes. This was the Chesapeake & Ohio Railroad's outlet for westbound and northbound coal from its Big Sandy, Logan, and Kanawha divisions. During its prime years, the yard was actually a series of seven yards, spread along a curve in the Ohio River Valley almost four miles long, paralleling U.S. 23 through four communities: Flatwoods, Russell, Raceland, and Worthington. The main

The Chesapeake & Ohio Railroad's Russell Yard, once the largest private railroad yard in the world. Chesapeake & Ohio Historical Society.

yard had thirty-six tracks and a seemingly endless supply of loaded coal hoppers. The twenty track coal classification area is a favorite spot of rail fans to watch for old coal hoppers bearing the insignias of the Baltimore & Ohio, Chesapeake & Ohio, Clinchfield, and Louisville & Nashville Railroads. The yard has an active switch tower, a retired Chesapeake & Ohio depot, and a restored Chesapeake & Ohio bay-window caboose.

The Russell Yard is still the largest CSX facility east of the Mississippi but now employs fewer than three hundred workers. In the 1950s the yard had almost four thousand men repairing locomotives and building cabooses, coal hoppers, tankers, and boxcars. Mile-long freight trains were routinely assembled and dispatched, and two roundhouses, two turntables, and a large locomotive shop handled some of the biggest steam engines of the day. The coal, freight, and passenger traffic was once so regular that locals remember being able to set their watches by the arrival and departure of the trains. At the edge of town, off U.S. 23, is the Rail City Hardware Store, where much of the area's railroading history has been preserved. Along the walls and shelves are model trains, railroad equipment, lanterns, and photographs of the Russell Yard during its prime.[6] Continuing south along U.S. 23 into Ashland, there are constant reminders of the importance of coal and iron in this area of the Ohio and Big Sandy Rivers.[7]

Ashland

Eastern Kentucky's largest city, Ashland (pop. 21,800), is located on the banks of the Ohio River. The area was originally known as the Poage Settlement, a pioneer community dominated by the Poage family of Virginia who came here in the 1780s. In 1854 members of the Poage family joined with Ohio investors and formed the Kentucky Iron, Coal, and Manufacturing Company to develop industry along the Big Sandy River. The new industrial center was named Ashland for Henry Clay's home in Lexington. Engineer Martin T. Hilton laid out the town's main streets, which were one hundred feet wide, parallel to the river; he then ran the cross streets, which were eighty feet wide, at right angles. His plan was to develop the city into a center of manufacturing, linking the iron furnace towns of Boyd County with markets in the Ohio Valley. The Ashland Iron & Coal Railroad and the Chattaroi Railroad further expanded the city by opening up the southern counties and the coal mines at Peach Orchard in the late nineteenth century.

Hilton's plans for the expansion of Ashland were a bit slow in coming, but by the 1920s the city had grown to a sizable regional center home to more than 29,000. By then, steel production had become the city's main industry. The American Rolling Mill Company (Armco) of Middletown, Ohio, bought the Ashland Iron and Mining Company in 1924 and the Norton Iron Works in 1928. By the 1930s they built a sheet rolling mill and, in 1953, a hot strip steel mill. In 1989 Armco became a partner of Kawasaki Steel, bringing a $350 million investment into the company. In 1999 it merged with AK Steel.

U.S. 23 is surrounded by industry through the Ashland area. The road is never far from the river or the rails and coal is everywhere. U.S. 23 passes by the Ashland Works steel mill, the AK Steel Corporation coke plant, and massive mountains of coal. Ashland's Kentucky Highlands Museum on Winchester Avenue gives an overview of the region's industrial development and culture, featuring archeological, social, and manufacturing artifacts.[8] Some of the permanent exhibits portray Native American Adena culture, rail, steam, coal, and industrial history.

Dominating the city are iron, coal, steel, and rail sites. The Ashland Furnace, once located at Winchester Avenue and Sixth Street, was built by the Ashland Coal and Iron Railway Company in 1869. It stood more than sixty feet high and had a capacity of forty tons of iron. Over the years it was expanded and rebuilt to hold more than five hundred tons. After 1921, it was operated by the Armco Steel Corporation and was known as the Sixth Street Furnace. Dismantled in 1962, it was the oldest known blast furnace in operation. The Norton Furnace, once located at Winchester Avenue and Twenty-third Street along U.S. 23, was built a few years later in 1873 and continued to operate well after the closing of the Sixth Street Furnace. Kentucky highway markers have been placed at both sites.[9] There are several other furnace sites near Ashland. The Clinton Furnace once stood near U.S. 60 and KY 538, just southwest of the city. George, Thomas, William, and Hugh Poage built this furnace in 1832, and for the next thirty years it produced up to fifteen hundred tons of iron annually. West of Ashland, near the Greenup County line, lies the ruins of the Princess Furnace, an iron-jacketed furnace, built in 1876 by Thomas W. Means (1803–90). Means was a leader in the iron industry for more than fifty years and owned furnaces throughout Kentucky, Ohio, Virginia, and Alabama. When charcoal became scarce in the 1870s, the Princess Furnace used stone coal as a fuel. These furnaces produced tons of pig iron that was later turned into wrought iron, steel, and ingot iron. Kentucky highway markers indicate both sites.[10] The nineteenth-century offices and company store of the Ashland Coal & Iron Railroad are located on Front Street and are listed on the National Register of Historic Places.

Many of the homes in central Ashland were built by wealthy industrialists. The Bath Avenue Historic District between Thirteenth and Seventeenth Streets is listed on the National Register of Historic Places and has many fine examples of late Victorian and Queen Anne architecture. John Means built the largest house of this era at 1420 Bath Avenue in Ashland's most prestigious neighborhood. Means was president of the Eastern Division of the Lexington and Big Sandy Railroad and was active in civic affairs, serving as Ashland's mayor in the 1880s. Brick pilasters accent the front and side of the house, and the property is surrounded by a fine iron fence. Hugh Means, director of the Kentucky Iron, Coal, and Manufacturing Company, lived at 1504 Bath Avenue, a home designed by Martin Hilton. Iron furnace financier Thomas Means lived at 1516 Bath Avenue, in a home that was later purchased by John C.C. Mayo's widow, Alice. She and her second husband, Dr. S. P. Fetter, moved into the house in 1917. It was originally built in the 1860s for Col. John F. Hager but underwent extensive remodeling in the twentieth century. Alice spent the first few years redecorating and refurbishing the home and lived here until her death in 1961. During the 1980s, the Mayo Manor served as the home of the Kentucky Highlands Museum.[11]

The residence at 1600 Bath was built by Ashland founder Robert Poage as a wedding present for his son. At 1520 Chestnut Drive, an Italianate-style house was built by iron industrialist Thomas Means as a gift for his daughter in 1876. An unusual home at 1600 Central Avenue was built in the Queen Anne style on Winchester Avenue and hauled by mules to its new location in 1919. This home's eclectic style includes cast-iron griffins to keep evil spirits at bay. City engineer Martin Hilton's residence at 1317 Hilton Court is an I-form house situated on the site of Indian burial mounds. Its overhanging roof, exterior chimneys, and single-story wing were added in 1900.[12]

While essentially a river town, Ashland is also the gateway to the east Kentucky mountains and is closely associated with Appalachian culture. The Jesse Stuart Foundation, on Winchester Avenue, is dedicated to "preserving the legacy of Jesse Stuart, W-Hollow, and the Appalachian way of life." The foundation is a regional press and bookseller that specializes in the works of east Kentucky author Jesse Stuart (1906–84). Stuart was born in a one-room cabin in W-Hollow, near Riverton, Kentucky. His father was a coal miner and tenant farmer. As a schoolteacher, Stuart made Greenup County his home and the site of most of his fictional writings. He captured the spirit and character of the mountains and produced more than fifty books and five hundred short stories during his career. Stuart has been described as a "farmer, schoolteacher, hunter, fisherman in Greenup, Carter, and Floyd Counties, [who] brings to his fiction ancestral links with the region, personal familiarity with places and history, sympathy with even at times a little scorn for their indigenous citizenry." In 1954 Stuart was named poet laureate of Kentucky and the next year received the Academy of American Poets Award. He returned to W-Hollow in the early 1960s and stayed there the rest of his life. His works have been celebrated around the world, and any one of his many collections of writings will give visitors to the Big Sandy region an appreciation of its people and culture.[13]

Catlettsburg

The CSX tracks, the Big Sandy River, and U.S. 23 lead south to Catlettsburg (pop. 1,960), established in 1798 and named for its first settler, Alexander Catlett. About 1803 the town sprang up around a tavern operated by Catlett's son Horatio. As steamboat trade developed along the Big Sandy River in the 1830s, the settlement became known as Mouth of Sandy. When a thriving timber trade developed on the river, Catlettsburg became one of the world's largest hardwood timber markets. Felled trees were dragged by oxen and mules to the creeks that feed the river, and large timber rafts were floated north to the town for sale in Ohio Valley markets. Its location at the mouth of the Big Sandy River allowed timber to be brought from both the Tug and Levisa Forks, as far south as Pike County. By the 1870s Catlettsburg was a booming town with hotels, an opera house, a race track, and endless saloons.

The railroads increased trade but also changed the character of the town. The Big Sandy Valley Railroad arrived in 1873; the Chattaroi Railway and the Lexington & Big Sandy Valley Railroad, in 1879. The building of the Chesapeake & Ohio Railroad led to the destruction of many older structures as its tracks cut through the town. The railroads have

A barge loaded with coal moves along the Big Sandy River at Catlettsburg, Kentucky.

continued to play an important role in the development of the city, and today Catlettsburg lies on an Amtrak route and is one the few cities in the area with a passenger depot.

Catlettsburg's heyday was during the late nineteenth century, when it had almost four thousand residents. After timber, coal was brought north along the Big Sandy but bypassed the town. Catlettsburg continued to prosper in the early twentieth century, but its economic base and population slowed. By midcentury, it was overshadowed by Ashland and nearby Huntington, West Virginia. The central commercial district of Catlettsburg has a number of structures from the late nineteenth century, but railroad expansion and a series of devastating fires in 1884, 1919, and 1932 have taken heavy tolls. The remaining commercial buildings are primarily two-story brick and masonry structures reflecting late Victorian commercial designs. Catlettsburg's downtown district has a distinct nineteenth-century atmosphere and serves as a great place to watch modern riverboats and barges transporting materials out of the Big Sandy and into the Ohio River, especially the many coal barges that can be seen along the banks. The "Scenic View" signs guide visitors through the floodwall to a small park that offers the best views.

Continuing up the Big Sandy River

From the Catlettsburg area U.S. 23 winds south along the Big Sandy and the former Chesapeake & Ohio tracks and leads right into the heart of the east Kentucky coal country. This highway can be followed all the way to Pikeville, where it merges with U.S. 119, allowing the option of continuing west toward Harlan County or heading east and crossing into West Virginia. U.S. 23 passes through rolling farmland as it enters the Big Sandy River Valley, and for the next thirty miles the railroad and the river are almost always in sight, just east of the highway. The terrain becomes mountainous farther south as the road leaves the rolling hills and cuts into the valley. South of Kavanaugh, U.S. 23 enters Lawrence County (pop. 15,569), and at Louisa (pop. 1,990), a regional commercial center, the river divides into its two main tributaries, the Levisa Fork and the Tug Fork. U.S. 23 continues southeast along the west bank of the Levisa Fork. The railroad tracks pass through Louisa and continue south along the east bank to Thelma.

Formed in 1821 from parts of Greenup and Floyd Counties, Lawrence County was named for Capt. James Lawrence of the USS *Chesapeake* in the War of 1812. Lawrence County has remained primarily rural throughout its history. During most of the nineteenth century, subsistence farming, hunting, and timber were the main sources of income for its residents. There was some limited iron work in Lawrence County, with Pioneer Furnace being the southernmost blast furnace in the Hanging Rock Iron Region. (A Kentucky highway marker indicates the former site of the furnace, just south of Louisa near KY 644.) The depth of the Big Sandy River in Lawrence County and the use of river transportation allowed for the limited development of timber and coal mining. Coal mining has been sporadic since the 1910s and Lawrence never developed many coal company towns or attracted the large coal companies. Its population has remained small during the twentieth century, and agriculture, particularly tobacco and livestock, has provided additional income and employment.

Like most people in the United States today, east Kentuckians get their electrical power from coal. In 1963 the Kentucky Power Company opened a large electric power plant north of Louisa. Utilizing the vast amounts of coal mined in the area, the plant supplies more than one million kilowatts of energy to customers throughout east Kentucky. The giant concrete tower cools 120,000 gallons of water per minute for steam condensing. It is more than three hundred feet in height and can be seen from miles away. This cooling tower was the first of its kind built in the Western Hemisphere, and a Kentucky highway marker details its history. Today coal trucks traveling along U.S. 23 provide a steady supply of coal for the plant.[14]

Richardson (formerly Peach Orchard)

Peach Orchard was located about fifteen miles south of Louisa on KY 1690 and can rightly claim the distinction of being the first coal town in the Big Sandy Valley. There were actually two Peach Orchards, designated as "old" and "new." The first of these communities was located along the banks of the Levisa Fork, just north of present-day Richardson (pop. 50). The second was located along Nat's Creek where the town of Richardson was built as a railhead for coal being produced at New Peach Orchard. Although this was a booming mining center in the late 1800s, only a few scattered buildings remain today.

Judge Archibald Borders (1798–1882), a pioneer farmer, merchant, and politician of Lawrence County, was responsible for the introduction of coal mining at Peach Orchard. Borders and his family ran a variety of small rural enterprises, including a saddlery, shoe shop, and timber business. He also built the *Big Sandy*, a river steamboat, and was instrumental in opening the area to merchant trade. Borders arranged for Cincinnati investors to develop mines along the Levisa Fork, selling them approximately two thousand acres of his land. The Peach Orchard Coal Company was formed in 1847 and, largely due to the efforts of Superintendent William B. Mellen, a town of white frame cottages was constructed during the 1850s from lumber cut by a company sawmill. The cottages housed the miners, many of whom were immigrants from the British Isles and Germany. Mellen also built a large mansion, a general store, and several public buildings. Mellen and Borders were visited in the 1850s by Elizabeth Haven Appleton, who produced one of the first detailed accounts of life in the Kentucky hills (and some of the first negative hillbilly stereotypes) through her writing for the *Atlantic Monthly*. In her fictional essays she described a primitive lifestyle portraying mountaineers as a crude, backward, and ill-mannered people. Nor was Appleton impressed by the habits and lifestyle of local elites. She noted soiled cutlery and crockery, a basic but hearty diet, and poor clothing.[15]

As coal was mined at Old Peach Orchard, it was shipped by barge along the Big Sandy River, to the Ohio River, and on to Cincinnati. The Civil War disrupted mining at Peach Orchard, and Mellen left his position to join the Union army. His assistant, Henry Danby, attempted to preserve company property as best he could, but little coal was actually mined during the war. The last barge load of coal left Peach Orchard in the spring of 1861. The Borders family remained, working with Union troops during the war and allowed Gen. James Garfield to use the *Big Sandy* to carry much-needed supplies to Pike County farther up the river. In the years following the war, the town was revived by George S. Richardson, who made the mines profitable again. Largely through his efforts, a standard gauge railroad was constructed from the Ohio River to Peach Orchard.[16]

The completion of the Chattaroi Railroad in 1882 helped begin a new era for mining. Calling themselves the Great Western Mining and Manufacturing Company, the owners opened two new mines along Nat's Creek and began the construction of a second, larger town. The Palace Hotel was the focal point of New Peach Orchard, and the new town, located only three miles from the older community, soon dwarfed Old Peach Orchard. By 1892 the Peach Orchard mines employed 556 miners who produced hundreds of thousands of bushels of coal each year. New Peach Orchard soon contained a church, school, large general store, railroad station, and post office and competed with Richardson as a flourishing center of trade.[17]

New Peach Orchard obtained a sophistication rarely matched by mining communities in Kentucky during the nineteenth century. Telephone service arrived in 1890 and streetlights were installed the following year. At the same time, new immigrants from Hungary and Italy began to join the workforce.[18] But the Peach Orchard area had a stormy history during the later years of the century. In September 1891 the mines at Old Peach Orchard ceased operations. Two years later the Knights of Labor led a month-long strike, disrupting production at New Peach Orchard. A national depression further damaged the Great Western Mining and Manufacturing Company. In 1894 the new Peach Orchard Coal Company took control of the mines but had limited success. Production began to decline even though the company opened a new mine and invested in new equipment. Despite these efforts, only forty men were employed at the mines by 1908.[19]

Sporadic attempts were made to revive mining along Nat's Creek after 1911, but most were small operations that lasted for short periods of time. After World War II, strip-mining operations began in Lawrence County. In 1970 a Texas firm began stripping along Nat's Creek at the site of Peach Orchard. In the process, most of what remained of both Old and New Peach Orchard was destroyed.[20] The Big Sandy Valley's first coal town was reclaimed by the land.

To reach the site of Peach Orchard, follow KY 1690 south from Louisa for twelve miles. The road approaches the Levisa Fork of the Big Sandy, the CSX tracks, and finally the settlement of Richardson (New Peach Orchard). One of the few remaining buildings at Richardson is the old company store, now a grocery. The wooden frame building is a two-story structure resembling a house, with a full front porch and series of second-story windows. A few scattered houses and some rotting foundations in the nearby woods are all that is left of the two communities today. A home built by Archibald Borders and his son David in the 1850s is located off KY 581. The antebellum two-story brick home is located on the hillside overlooking the Levisa Fork and is surrounded by rolling

One of the last buildings remaining from the Peach Orchard coal town, Richardson, Kentucky.

Coal Miners at Peach Orchard, Kentucky, circa 1890. Courtesy of Appalachian Photographic Archives, Alice Lloyd College.

farmland. On the grounds of the farm is a family cemetery where a ten-foot spire-type gravestone resting upon a square pedestal marks the burial site of Judge Archibald Borders. His wife, Jane, and son David are also buried here.[21]

Southeast of Lawrence County, across the Levisa Fork, lies Martin County (pop. 12,578), where, despite a small population, coal is an important part of the economy. There are few coal towns, but several large underground mines and extensive strip-mining provide income and employment for the area. The county was formed in 1870 from portions of Floyd, Johnson, Pike, and Lawrence Counties and was named for Congressman John P. Martin of Prestonsburg (1811–62). Early nineteenth-century settlement appeared along the Tug Fork of the Big Sandy River, and by the late 1860s Warfield, the first county seat, had developed into a small regional center for coal and salt mining. During the Civil War some of the area coal mines were used to hide from marauding enemy troops. In the late nineteenth century, Warfield native Lewis Dempsey, a local financier and developer, acquired vast tracts of land that were leased to coal and timber companies. During the twentieth century, Martin County lagged far behind other areas of the Big Sandy River Valley because the Norfolk & Western Railway did not build a spur connecting the county with its main line in West Virginia until the 1970s. In the 1990s the Martin Coal Corporation operated one of the largest surface mines in the country, employing more than a quarter of the county's workers. Limited timber, oil, and natural gas complement the county economy.[22] Despite limited development until the late twentieth century, Martin County was the site of one of the region's unique coal company towns, the settlement of Himlerville, today called Beauty.

Beauty (formerly Himlerville)

The small community of Beauty (pop. 600) is thirty-five miles southeast of Louisa on KY 40. It can be reached by following U.S. 23 to KY 645, just before Ulysses. At Inez, KY 40 continues east to Beauty. It can also be reached from West Virginia by crossing the Tug Fork near Kermit. Beauty was originally known as Himlerville, named after utopian socialist Martin Himler, a Hungarian Jew who devoted most of his life to the cause of immigrant mine workers and their plight. Today, Beauty is a small community in a remote part of Martin County but has several significant structures and an interesting cemetery that make it worth a visit.

Moved by the exploitation of his countrymen in the coalfields of Appalachia, Himler attempted a bold experiment in cooperative mining. His dream was to provide good living and working conditions for miners and to allow them a voice in company policies and practices. While acting as editor of the *Hungarian Miners' Journal,* Himler founded the Himler Coal Company. He then offered stock exclusively to Hungarians who would serve as workers once the company began to develop mines. Himler himself owned less than 3 percent of the stock, with the remaining shares divided among fifteen hundred stockholders.[23] With the pool of money raised from Hungarian immigrants, the Himler Coal Company began operating in West Virginia in 1916. The following year, at the peak of wartime production, it purchased coal leases on land west of the Tug Fork in Martin County, Kentucky. In order to market the coal mined on the Kentucky side of the river, Himler had to invest a significant part of the company's capital in constructing a railroad bridge linking Kermit, West Virginia, to Warfield, Kentucky. This steel and concrete bridge allowed the

Norfolk & Western Railway access to the Martin County coalfields and continues to be an important transportation link in the area. Although it was originally estimated to cost $25,000, the company spent more than $300,000 by the time it was completed. Despite the financial setback, the first train reached Himlerville on 21 May 1921, amid much celebration.

The Himler Coal Company soon built a large tipple, a power plant to provide electricity, and high-quality, well-constructed houses for its workers. Housing plans were similar to those used by Consolidation Coal at Jenkins, a model company town constructed a few years earlier. Each five-room home featured gas and electricity, fireplaces, tubs, showers, and garden areas. Miners could purchase the home or have one built to their liking. A company headquarters was constructed and served as both an office and general store. Unlike most companies, it did not pay its workers in scrip. The town also had a clubhouse with a library, auditorium, printing shop, and a Hungarian bakery. Hungarian dances were held, and Himlerville's school even held classes in Hungarian culture.

Himlerville attracted immigrant miners from other coal towns. The family of Steve Balazs came from Stonega, Virginia, in 1921. Balazs's father had been a coal miner in Hungary and made the journey to the States before the war. Balazs recalled few native white workers in the town but a large number of Hungarians, Slavs, and Germans who found Himlerville a pleasant place to live. Himler Coal organized recreational activities, a brass band, and sporting teams and built an opera house that featured entertainers from New York City and Philadelphia.[24]

Like many other coal operators, Himler had a paternal attitude toward his employees and enforced a strict code of conduct in the town. But Himler, unlike some other owners, was motivated by a genuine desire to help his miners and to maintain a sense of solidarity among workers. Any vices that threatened the order of the community were dealt with immediately. Miners caught gambling were discharged and their shares of stock were repurchased by the company. When Himler found a small store dealing in contraband whiskey, he promptly discharged the owner, and the company purchased the store and house properties. Strict social control and a high standard of living gained Himlerville an enviable reputation. A survey of hundreds of coal towns in the 1920s conducted by the U.S. Coal Commission rated Himlerville second in living conditions.[25]

Despite its well-deserved reputation, by the mid-1920s Himlerville's days were numbered. The company had greatly underestimated the costs of building the town and equipping the mines. It was just beginning to make a profit when the wartime demand for coal slowed. Although production at Himlerville reached its peak from 1924 to 1926, the falling price of coal during the 1920s forced an end to Himler's experiment. Despite the workers' vote to substantially reduce wages in 1928, the Martin-Himler State Bank failed and the entire company soon collapsed. To make matters worse, an explosion in the mines killed several miners on 28 June 1928, and that evening a flash flood destroyed much of the town. Himlerville never recovered.[26] In 1929 the assets of the Himler Coal Company were purchased at public auction by Arch Hewitt of Huntington, West Virginia, who later formed the Martin County Coal Corporation. Although mining operations continued and the town retained a population of three hundred to four hundred people, Himlerville lost its unique flavor as an island of Hungarian culture in Appalachia.[27] It also had difficulty competing in coal markets. Under Himler, the quality of coal mined was not always up to market standards. J. H. Mandt, who operated some of the mines in the 1930s, claimed that he had trouble marketing coal from Himlerville because the company had a reputation for producing dirty coal. To disassociate his business, he had the town renamed Beauty and marketed coal under that name.[28]

A few of Himlerville's prime structures remain today. In the center of town along KY 40 is a two-story brick and masonry building that was

Martin Himler *(on platform in miner's cap)* addresses a stockholders' meeting at Himlerville, Kentucky. Reprinted from *Coal Age* (1921).

Himler's house is visible on the hillside above the town.

once the Himler Coal Company headquarters. It is now a repair shop. On the hill above the headquarters is Himler's house, accessible by a long flight of concrete steps. It is now a private residence. The tree-shaded lot hides all but the steep-pitched front gable and second floor of the home, which are visible from street level. The spacious home overlooks the town and has been placed on the National Register of Historic Places. Below the house and behind the repair shop are a few remaining miners' houses. They are basic one-story wood-frame, three- and four-room dwellings that have not been well maintained over the years. A short distance to the right of the headquarters is the former Martin-Himler State Bank, a one-story native stone building, with arched window bays and lathed wooden railings, that has been painted white. In the wooded area directly across from town is the old Hungarian cemetery, a resting place for Himlerville miners. Each grave is surrounded by an iron fence and features tombstones with inscriptions in Magyar. The cemetery lies on private property and is neither visible nor accessible from the road.

From Lawrence County, U.S. 23 continues south and remains west of the river and railroad as it enters Johnson County (pop. 23,445), which was formed in 1843 from portions of Floyd, Lawrence, and Martin Counties and named for Richard M. Johnson (1781–1850), the hero of the 1813 Battle of the Thames and vice president of the United States (1836–40). Some coal mining was developed in Johnson County prior to the Civil War, but it was not until the 1880s, when railroads entered the area, that it became an important part of the local economy. By the early twentieth century, coal mining replaced farming as the area's principal economic activity, partly because of the efforts of Paintsville resident John C.C. Mayo. He helped develop a coal industry along Miller's Creek and transformed Paintsville into a thriving regional commercial and cultural center.[29]

Paintsville

Paintsville (pop. 4,132) is located on KY 321 (old U.S. 23) where it meets U.S. 460.[30] It derives its name from Paint Creek, an area where early settlers were fascinated by the Indian drawings on tree trunks along its banks. A trading post known as Paint Lick Station may have been built at the city's present site, but a town did not emerge until Rev. Henry Dickson (Dixon) of North Carolina gained possession of the land in the early nineteenth century.[31] The Civil War disrupted the town's economic development as Union and Confederate troops fought for control of the Big Sandy. Paintsville's most dramatic period of growth began in the early twentieth century when the city emerged as the center of a vast coal mining region. Paintsville entrepreneur John C.C. Mayo (1864–1914) amassed a great fortune by purchasing the mineral rights to coal lands throughout the Big Sandy Valley and by attracting large corporate investors to the area. Mayo made his home in Paintsville and, in an effort to improve the cultural life of the town, he constructed a mansion and a new Methodist church and helped found a seminary.[32] The coal industry brought new prosperity and many new residents to Paintsville, but it also created new social problems. The influx of foreign immigrants and African Americans into nearby coal towns led to a nativist backlash in the 1920s, including activities by the Ku Klux Klan.[33]

Paintsville has adapted to the economic changes of the late twentieth century much better than other communities by attracting new industries and developing recreational sites and tourist attractions. In 1980 the U.S. Army Corp of Engineers constructed an earthen dam across Little Paint Creek, four miles west of the city on KY 40, and created Paintsville Lake, which is now used for fishing, boating, swimming, and hunting. Three Paintsville sites are related to the life of John C.C. Mayo and are located in the center of town at Third and Court Streets. The first, the Mayo Mansion, was constructed between 1905 and 1912 at a cost of more than $250,000. Mayo and his family lived in this beautiful home until his death in 1914. The mansion is located on the corner of the two streets, with a native stone wall surrounding the spacious grounds. The three-story house is made of brick and stone and features a series of dormers and chimneys. Steps lead up to the entrance of the mansion and a row of two-story-high white columns spread across the front of the building. After John Mayo's death, his widow, Alice, donated the building to the Big Sandy Valley Seminary. In 1945 it was sold to the Catholic church and has served as Our Lady of the Mountain Catholic School for almost sixty years.

Next to the mansion is the Mayo Methodist Church, the largest church in Paintsville, reaching more than three-stories high. This late Gothic revival church is made of native stone and features impressive windows. Fellow industrial capitalist Andrew Carnegie, a friend of Mayo, donated a pipe organ to the congregation. On the other side of the mansion, on the

The beautiful grounds of the Mayo Mansion in Paintsville, Kentucky.

Coal baron John C.C. Mayo's private office on the grounds of the Mayo Mansion in Paintsville, Kentucky.

grounds of the school, is Mayo's former one-story brick study, where he conducted business transactions and met with visitors. Today it is used as an office and residence of the principal of Our Lady of the Mountain School. The mansion, the Mayo Methodist Church, and the study are all listed on the National Register of Historic Places, and Kentucky highway markers, located in the front lot, describe the life of Mayo.[34]

Along the remainder of Third Street, next to Mayo's office, is the site of the old Methodist seminary established by Mayo, which later evolved into John C.C. Mayo College. Due to financial problems, the college closed in 1928. Later efforts to transform the school into a junior college operated by the University of Kentucky failed. However, oilman Everett J. Evans began a movement in the late 1930s to create a "practical college" in Paintsville, and in January 1938 the legislature responded to local efforts by approving $56,000 for the creation of the Mayo State Vocational School. The vocational school originated as a response to the growth of the mining industry. Thousands attended the ceremony marking the transfer of the Mayo estate property to the state of Kentucky on 26 July 1938.[35] Until recently the vocational school continued to offer job training programs to area residents and employed a faculty of fifty instructors.

In an effort to preserve the memory of rural mountain life in Appalachia before the coal boom of the early twentieth century, the Paintsville Lake Historical Association created the Mountain Home Place at the dam site. It is both a working mid-nineteenth-century farm and a place to preserve historic structures displaced by the creation of the lake. Central to the Mountain Home Place is the one-and-one-half-story McKenzie log cabin. Built in 1861 by David McKenzie, it is constructed of yellow poplar held together with half dovetail notches. Costumed guides perform farm chores and fieldwork, animals roam the grounds, and traditional wooden fences line the property. A small wooden schoolhouse, with simple benches and a coal stove, is a popular sight at the Home Place.[36] The best time to visit Paintsville is during the Kentucky Apple Festival held the first weekend in October. This weeklong celebration, which began in 1962, has evolved into one of the largest fairs in the state. The festival culminates on Saturday night with a performance by a major country music star in the Johnson County High School Gym.

Van Lear

The largest mining community in Johnson County during the early twentieth century, Van Lear (pop. 1,050) became one of two model company towns constructed by the Consolidation Coal Company in east Kentucky. Like its sister town of Jenkins, Van Lear developed into a community in the truest sense of the word. Retired miners and their descendants continue to preserve the rich heritage of their community in the pages of *The Bankmule,* the official publication of the Van Lear Historical Society. In recent years, the town has developed a growing tourist industry as the hometown of country music star Loretta Lynn, known as the "coal miner's daughter." Located along both banks of Miller's Creek about four miles south of Paintsville on KY 302, Van Lear has many structures remaining from the early twentieth century and preserves the atmosphere of a booming coal town.

The development of Van Lear was closely tied to the career of Appalachian financier John C.C. Mayo. In 1887, after attending Kentucky Wesleyan College, Mayo returned to Johnson County and took a job teaching school in the Miller's Creek area for forty dollars a month.[37] At the time, Miller's Creek was undergoing a boom in timber trade and was better known to lumberjacks than coal barons. In 1886 the creek powered three sawmills. Mayo saw the potential for coal mining in the wooded valley and purchased the mineral rights. In 1906 he began a mining operation along Miller's Creek within walking distance of the school. To publicize the quality of his coal, Mayo shipped several large blocks to Jamestown, Virginia, where they were used to construct a coal house in an exposition marking the tercentenary of the founding of Virginia. In 1909 he sold his Miller's Creek mineral rights to the Consolidation Coal Company of Baltimore, Maryland. That same year, the Miller's Creek Railroad, a trunk line, linked the Miller's Creek coalfields with the Chesapeake & Ohio Railroad at a site now known as West Van Lear. These developments set the stage for the rapid construction of a large coal town.[38]

The new community, built between 1909 and 1914, was named for Van Lear Black, the director of the Consolidation Coal Company. Among the more interesting features was a steam-powered electric plant and a water processing plant. By 1910 Consolidation Coal was able to produce electricity for every mine, house, and business in the town. (In 1922, the larger power plant constructed at Jenkins began transmitting power to the town and the Van Lear facility was closed.) A large three-story office complex was constructed, as well as a recreation building, four schools, a hotel, and seven hundred houses. By the 1920s five mines were operating along the Chesapeake & Ohio tracks at Van Lear, each mine having its own preparation plant and tipple. The town had grown from a population of 337 residents in 1912 to more than 4,000 by 1920, making it the largest town in Johnson County.[39]

From the beginning, Consolidation Coal intended Van Lear to be a model company town. The first meeting of the town's board of trustees was held in the company's supply house, where regulations were developed for the people living in Van Lear. Among laws passed by the town fathers was one imposing a fine of five to ten dollars on any minor found with cigarettes in his possession. Ten-dollar fines were to be levied against anyone who abused an animal, rode or drove more than ten miles an hour, or congregated "upon the streets or in houses for any immoral or unlawful purpose."[40] C. Mitchell Hall described Van Lear in the 1920s as a "modern little town. Being controlled by one of the largest coal corporations in the United States, built and maintained by trained and technical men from the time of its creation, it has developed into a type of town where living conditions are such as to be inviting to anyone, with good houses, good streets and roads, water system, schools, churches, hotels, clubhouse, picture show, ball park, and recreational grounds."[41]

Consolidation Coal recruited its workforce from a variety of sources. Most of the managers were brought in from older operations in Pennsylvania, Maryland, and West Virginia, and they occupied houses along Van Lear's "Silk Stocking Row," a section of high-quality homes. Common miners were chosen from a large pool of local farmers, immigrants, and African Americans. Van Lear had a relatively small African American population, and like all Kentucky school districts of the day, the Van Lear city schools were segregated. African American children had to attend the district's colored school, a building that also served as a church and lodge for the black community. Far more numerous were immigrant workers, including Austrians, French, Germans, Irish, Italians, Lithuanians, Poles, Russians, and Slavs, who all found work at the coal mines in Van Lear.[42] Unlike native Appalachians, many of these immigrant workers were Catholic, and on 3 December 1911 Catholic residents dedicated Saint Casimir Church.

Although there were occasionally tensions among the different ethnic groups in Johnson County, there were few problems in Van Lear. Public officials moved quickly to preserve harmony in their town. In September 1915, the town board passed an ordinance making it illegal to "publicly denounce, abuse, or use any insulting language against any religious denomination or sect of people, or to use any language intending to disturb or insult any person or persons, or calculated to provoke an assault."[43] In contrast to the intolerance in the area, Van Lear was an island of civility. When interviewed, many of the people who lived in the town in the 1920s and 1930s expressed fond memories of their immigrant neighbors. Henry Skaggs recalled attending the funeral of George Sotnikoff, the son of Russian immigrants. Skaggs regarded Sotnikoff as one of his best friends.[44] When local writer Joyce Meade asked residents about immigrants Frank and Mabel Campigotto, their responses were almost universally positive. They were characterized as "really a good family," "people who cared," "wonderful neighbors," "they don't come no better," and, last but not least, "he made the best spaghetti and meatballs you could ever eat!"[45] Charlene Conley noted that one of the staunchest supporters of the Van Lear High School football team during the 1930s was Italian immigrant Andrea Marino, who often helped out the team.[46]

In Van Lear, miners of many different origins shared the experiences of mining and life in a company town. There was plenty of work available, and the town thrived. But coal mining was also a dangerous profession, and Van Lear had its share of accidents and tragedies. Bruce Cantrell, who worked in the mines for twenty-seven years, "suffered a broken shoulder, broken leg in three places, and a broken back." And of course in the early twentieth century coal dust in the air took its toll on many miners who developed black lung. Overall, the Van Lear mines had a very good safety record, but on 17 July 1935 the No. 155 mine was rocked by a methane gas explosion. Fortunately, the mine was not working that day and only small maintenance crews were underground at the time. The men were a mile and a half from the entrance of the mine, and all nine were killed in the explosion. It took rescue crews more than twenty hours to break through the rubble and recover the bodies.[47]

The peak of activity in Van Lear was the late 1910s and 1920s. The Great Depression of the 1930s created problems for Consolidation Coal and it began to cut costs. The era of the great company coal towns was coming to an end. In 1946 Consolidation Coal sold its properties in Van Lear, and tenants were given the opportunity to purchase the remaining 247 houses in the town. Mine No. 155 was the last operated by Consolidation, closing in 1949; the tipple and offices closed one month later. The company leased coal lands to the Farwest Coal Company, who continued to operate mines in Van Lear until 1955.[48]

Van Lear is a very good example of a model company town of the Big Sandy Valley during the early years of the twentieth century. The landscape contains numerous structures from the mining era, including uniform company houses, company buildings, and abandoned mine sites. The winding contours of the landscape are typical of the region, with a railroad line through the center of town and houses tucked into the many hollows that run off of Miller's Creek.

Van Lear can be reached by driving south from Paintsville along KY 321 (formerly U.S. 23) to KY 1107. About one mile along KY 1107, at the junction of KY 302, is the former site of the steam-powered electric plant and water processing plant. Up toward the top of the hill are two water tanks from the old plant still standing as mute evidence of the town's former utility system. KY 302 continues into the heart of Van Lear, about one mile farther. Just past the S curve leading into town is a gas station and market on the left. This family-owned grocery is located in the building that once served as the engine repair shop for the Consolidation Coal Company.

The road's second S curve leads to Miller's Creek Road. Ahead is a large blue building marked "City Hall." This once served as Consolidation Coal Company's office complex. Today, the three-story building, with its pyramidal roof, dormers, and front steps leading to the second level, houses the Van Lear Historical Society, a grassroots organization collecting and preserving artifacts from the town's rich mining past. It contains displays of mining implements, a doctor's office, a store scene, and a scale model of Van Lear as it existed as a company town. The S curve continues past the city hall to Schoolhouse Hollow on the left. The apartment building with the address 300 once served as the town's high school. The public school gym is at the end of this short road. The construction of this large, beautiful native-stone building on the hillside was carried out as a WPA project from 1936 to 1938. It now serves as the local Masonic Lodge.[49] One-half mile farther along Miller's Creek Road is a row of houses along the right side of the road, the remaining portion of Silk Stocking Row, Van Lear's premier residences.

The immigrant community of Van Lear was relatively small during the early twentieth century, but it was an important influence in the town. Although census records indicate small numbers of aliens living in the county in the 1910s, their descendants maintained some important cultural traditions. St. Casmir's Church was once the most visible evidence of Van Lear's immigrant community. In 1910, Consolidation Coal donated a small lot for the construction of a Catholic church. The Catholic bishop sent Father Francis B. Sokol, a Polish immigrant, as pastor to the new congregation. The new parish was named Saint Casimir, in honor of the patron saint of Sokol's native land. He noted in 1913 that the company was very anxious to recruit more Poles because of their reputation as good miners. Father Sokol served the community until 1919 when he departed to serve in the Polish army. Other priests followed in his place, including Father Joseph Beruatto, an Italian immigrant, and Father George Metzler, a Frenchman. Unfortunately, the church was sold in 1941 and torn down in 1958, but the old Catholic cemetery and the rectory (or "the old priest's house," as it is referred to locally) remain. Jeanette Knowles transcribed church records, listing fifty-five names of individuals buried at the site. She also discovered six additional graves

Consolidation Coal's No. 155 Mine Store at Van Lear, Kentucky.

not listed in the records of the church. The old cemetery is located to the left of the house at 1178 Miller's Creek Road.[50]

Stone steps lead up the elevated lot to the two-story white wood frame rectory located five houses past the old cemetery. A small family cemetery containing the grave of Abe Goble is next to the priest house, though the Goble grave is not related to the story of the church. Local legend has it that Father Charles Donovan, the last pastor of St. Casimir, was cursed by his former housekeeper. The good father then began to witness strange occurrences, including the movement of inanimate objects by supernatural means. Trouble persisted and in 1941 a new mission center was built in Paintsville. Father Donovan soon relocated.[51]

Less than a mile farther on Miller's Creek Road is Webb's Stop and Shop, a site listed on the National Register of Historic Places. Built in 1918, this building once served as the company store for the workers at mine 155 and is affectionately known as "old number five." It is a two-story wood frame building set on a concrete and stone foundation and appears larger than it actually is because of its false-front facade. Inside, the layout of the store has not changed much in the past eighty years. The grocery is owned by Herman Webb, brother of country singers Loretta Lynn and Crystal Gayle, who maintains the singers' birthplace. Loretta Lynn was born Loretta Webb on 14 April 1935. She was raised in Butcher Hollow and her father worked in the Van Lear coal mines. After marrying Oliver Lynn, she left the area and began a career as a country singer on the West Coast in the state of Washington. She gained fame during the 1960s through her appearances on the *Grand Ole Opry.* By 1990 she had recorded fifty albums and received a Grammy Award. The 1981 autobiographical feature film *Coal Miner's Daughter,* starring Sissy Spacek and Tommy Lee Jones, depicts Lynn's life in Butcher Hollow, although it was filmed in other east Kentucky and southwest Virginia locations.

Butcher Hollow is located farther east on Miller's Creek Road. The pavement ends and a graded road leads to a fork. The left fork leads to the hollow, the home of Loretta Lynn and birthplace of Crystal Gayle. Past the entrance to the hollow, there is an abandoned large yellow wooden building on the right side of the road. This one-room schoolhouse was where Lynn attended classes in her youth, and was part of the Johnson County School District, not part of Van Lear. Less than a mile up the hollow is the Webb home place. It is well marked, but the site is privately owned. The structure is a large, ramshackle cabin of rough-hewn

The Webb Home in Butcher Hollow.

wood, with a chimney and full porch that sits on stilts and spreads across the front of the dwelling. Electricity and running water came late to Butcher Hollow, so residents used coal oil lamps and drew water from the nearby well. The interior of the home has been restored to show how the Webb family lived in the 1940s, and the Webb home place is reminiscent of how people lived beyond the model company town of Van Lear. Conditions in the rural areas of Johnson County remained quite primitive well into the mid-twentieth century.

Whitehouse

Whitehouse (pop. 150) lies along KY 3390 about eight miles northeast of Paintsville. To reach the remains of Whitehouse, follow KY 40 about ten miles east from Paintsville, across Two Mile Hill, and turn north onto KY 3390. At the end of KY 3390 are a few scattered houses close to the railroad tracks in this narrow valley. There is little left of old Whitehouse, as the few houses in the town were built after the coal mining boom. However, one can certainly get a feeling for the remote nature of this mining community.

The development of this coal town is closely linked to that of New Peach Orchard. In 1887 the Chattaroi Railroad was extended from Richardson to Whitehouse in order to exploit deposits of coal located there. Two Ohio-based companies quickly began large-scale mining operations, and by the end of the year the first shipments of coal were leaving Whitehouse.[52] Unlike those in the better capitalized mining towns of Peach Orchard and Van Lear, the companies operating out of Whitehouse invested little money in the construction of the town. Miners' houses were shacks with canvas or paper covering the interior walls. Dust from the mines and soot from the trains gave the community a grim and dirty appearance. From 1903 to 1907 the Whitehouse Cannel Coal Company even resisted giving its mines an appropriate ventilation system.[53]

In addition to mining, Whitehouse served for a time as the railroad's end of the line. Goods being shipped south had to be transferred to boats or wagons in order to continue the journey up Levisa Fork. Passengers were forced to disembark and continue over the rugged mountain terrain in small horse-drawn hacks. The Chesapeake & Ohio Railroad later purchased the line and extended its tracks southward in the early twentieth century.[54]

Though Whitehouse lost its position as a transfer point, significant physical improvements were made when the Northeast Coal Company assumed control of the town about 1910. New homes were constructed and far greater attention was paid to sanitary conditions. Mines were also opened in nearby Offutt, River, and Williamsport. (Northeast later developed the coal communities of Thealka and Auxier.) Unlike competing companies, Northeast only employed native, white Americans. Under the direction of Henry LeViers, Northeast implemented a paternalistic system within its town. Northeast closed its last mine at Whitehouse in 1931.[55]

South of Paintsville, U.S. 23 heads west away from the railroad and the river, enters Floyd County (pop. 42,441), and continues about nine miles to the county seat, Prestonsburg. This is a new stretch of U.S. 23 that bypasses many sites. A better way to enter the rich coal country of Floyd County is to follow KY 321 (old U.S. 23) south and remain along the railroads tracks and the river.

Floyd County was formed by an act of the Kentucky General Assembly in 1799 and named for Kentucky pioneer John Floyd. The original county was composed of the entire Big Sandy Valley and much of east Kentucky. Fifteen counties can trace their heritage to old Floyd County, including Floyd, Pike, Martin, Knott, Magoffin, and Johnson. During the nineteenth century, Floyd County was primarily composed of small farmers, many living in the areas of Middle Creek and John's Creek. The Chesapeake & Ohio Railroad penetrated Floyd County in 1903 and began constructing numerous rail spurs reaching into previously isolated areas. Coal towns soon appeared, and by the 1910s Floyd County was a major mining region.

Auxier

Auxier (pop. 900) lies south of Paintsville on KY 3051, just off KY 321. The site was settled in the late eighteenth century by members of the Auxier family. The location on the Big Sandy River, only one-quarter mile from the mouth of John's Creek, allowed the settlement to develop as a small trading center known as Hager Shoals. Push boats and occasional steamboats carried on trade with three general stores in the area. In 1904 the Chesapeake & Ohio Railroad purchased lands from the Auxier family, and the next year a C & O stop called Auxier was established. In 1909 a post office opened and the town was officially named in honor of James W. Auxier, a local resident. That same year, John C.C. Mayo purchased lands from Henry Litteral and leased the lands to the Northeast Coal Company.

Between 1910 and 1921 Northeast built most of the houses in Auxier and developed the town. A tipple was constructed along the C & O tracks and a bridge was built across the Big Sandy to serve motor vehicles. The town was complete by 1921, but a few more houses were constructed in 1927. The operators at Northeast employed only white, native-born workers, so the town never developed an African American or immigrant community. Auxier prospered under Northeast until the 1940s. Although several other companies operated in the area, Auxier began its decline in 1946 when Northeast closed its mines. By 1952 Northeast had sold most of its properties in Auxier.

Despite several attempts to develop manufacturing, including a clothing factory in the 1970s, Auxier has remained a quiet residential community. The town is interesting because it has several streets of original company houses and most are still occupied today. They are one- and two-story shotgun wood frame homes that line the two main streets of the town. The Auxier Methodist Church, a white wood frame building with a small cupola, was built in 1913, and the pews were donated by the

Houses and church built by the Northeast Coal Company at Auxier, Kentucky.

Mayo Methodist Church in Paintsville. The site of one of the Auxier mines now serves as the entrance to Jenny Wiley State Park.

Prestonburg

Prestonsburg (pop. 3,612), an important regional commercial center, lies at the east end of the Mountain Parkway and has served as the political heart of Kentucky's Big Sandy Valley since the early nineteenth century. John Spurlock is credited with building the first cabin on the site of the future city in 1791. The town was not surveyed until 1797, when agents for Col. John Preston directed John Graham to lay out a station.[56] Farmers throughout the region soon patronized Prestonsburg stores to obtain cloth, manufactured goods, and luxury items. By the 1820s Floyd County's seat could boast of a brick courthouse, and merchants were sending hundreds of pelts up the Levisa Fork in push boats for resale in distant markets.[57]

Prestonsburg is primarily a commercial river town, but it has long been associated with the coal industry. Samuel May (1783–1851) sold Robert Deering a tract of land in 1841 on which Deering planned to mine coal and grind corn. Deering's ambitious plans amounted to little, but later in the decade May signed several agreements relating to the production of wooden planks and the mining of coal. May's interest in lumber and coal was closely tied to the arrival of the first steamboats on the Levisa Fork. May left to join the California Gold Rush in 1849, but by then Cincinnati capitalists had become aware of the rich coal lands in the area.[58] In 1848 Archibald Miles of Cincinnati attempted to create a community south of Prestonsburg based on coal mining and milling. A similar venture was attempted by the Northern Kentucky Coal Mining Company west of Prestonsburg. Both companies failed to turn a profit, and the latter sold out to the Peach Orchard Coal Mines.[59]

The Civil War disrupted riverboat traffic, but the postwar years saw more steamboats and an expanded timber business. By the 1870s packet lines began to offer regular service to the people of Levisa Fork, giving residents of the area the ability to travel in comfort from Pikeville to Catlettsburg or Huntington. One company, the White Collar Line, made an arrangement with the Chesapeake & Ohio Railroad to facilitate the transfer of passengers and merchandise.[60] Though preferable to a push boat, a steamboat trip was long and slow, taking fourteen hours to travel from Prestonsburg to Pikeville.[61]

The town was transformed in 1904 when the track line of the Chesapeake & Ohio Railroad reached West Prestonsburg. A regular depot was constructed and a train stop was created across the river from downtown Prestonsburg. Passengers wishing to board the train were forced to cross a toll bridge and pay a five-cent fee.[62] The coming of the C & O rail line brought an expansion of the coal industry. The Middle Creek Coal Company and the Blue Beaver Coal Company began operating mines in West Prestonsburg. The Prestonsburg Coal Company began mining along Town Branch in 1908, and the Colonial Coal and Coke Company operated mines on both sides of the Levisa Fork, just to the south of Town Branch. Many of the miners lived in Prestonsburg and crossed the river in order to work.

The mines operated by Colonial Coal remain the most vivid link between Prestonsburg and the coal industry, even though Colonial Hollow never developed into a large residential area. Colonial Coal and Coke Company built the site in 1911, but at its peak the coal camp contained only thirteen single-family dwellings and a boarding house. However, the hollow also contained a large office/store complex and a coal tipple. The operation is best remembered for the aerial tramway that crossed the river in order to bring buckets of coal from the Prestonsburg side of the river to the tipple at Colonial Hollow. Mining boomed during the 1920s, but the Colonial mines closed during the Great Depression.[63]

Prestonsburg's best kept secret is the Colonial Hollow ghost town. Following the closing of the mines, the old road to Colonial Hollow was abandoned. Because of its remote location, the ruins of the community are preserved among the trees and bushes of the hollow. Colonial Hollow is located along the C & O tracks on the west side of the Town Branch Bridge. There are paths leading to the ruins of the collection point, the miner's bathhouse (clearly visible on the side of the mountain), and the foundations of several houses and the office/store complex. Most of Colonial Hollow is overgrown by vegetation during the summer months.

In 1929 construction began on three bridges designed to span the Levisa Fork at Prestonsburg. The first crossed the river at Cliff as part of the new Mayo Trail highway. The second gave Prestonsburg residents access to Town Branch, and the third linked the city to the rail depot at West Prestonsburg. The first was replaced in 1987, but the West Prestonsburg Bridge and the Town Branch Bridge have been placed on the National Register of Historic Places. These reinforced, concrete arch bridges have become a popular local symbol of Prestonsburg used in tourist and promotional literature.[64]

Prestonsburg has several other interesting sites to visit. The city emerged as a regional tourism center in 1954 when Jenny Wiley State Park was created around the reservoir formed by Dewey Dam. In addition to providing facilities for camping, boating, fishing, and swimming, the park boasts a sky lift and the May Lodge. Downtown Prestonsburg is typical of the regional commercial centers described throughout this guidebook. The Front Street Historic District, along Front Street between West Court Street and Ford Street, is listed on the National Register of Historic Places and has many commercial buildings from the early decades of the twentieth century when Prestonsburg was thriving. There are a variety of brick, stone, and masonry structures, most two stories high, that have served as stores, hotels, lodges, and offices. Arnold Avenue has many fine homes from the same era that have also been added to the National Register and includes examples of Queen Anne, Spanish revival, craftsman, classical revival, and late Victorian architecture.

Another site of interest is the Samuel May house on North Lake Drive, also listed on the National Register. This two-story, red brick, Federal-style house was built by Samuel May in 1817 and served as a center for political activity throughout the nineteenth century. During the summer of 1863 it was used by Col. Andrew Jackson May as a recruitment center while he formed the Confederate Tenth Kentucky Cavalry. The May house has also been active in developing exhibits, Web sites, and collections that pertain to the county's local history, including some of its coal mining towns, especially Wheelwright.

Most of Floyd County's coal towns are located south of Prestonsburg and can be reached from KY 80, KY 7, or KY 122. At Martin, still an important railroad marshaling yard for many area coal mines, the CSX tracks divide. One line continues south along the right fork of Beaver Creek to Garrett and Wayland and the other follows the left fork toward Drift, McDowell, and Wheelwright. These extensions of the Chesapeake & Ohio Railroad were completed in 1918 and opened up the southern part of the county to mining. KY 122 and KY 7 both junction with KY 80, a four-lane highway that connects Prestonsburg with Hazard. Travelers can use Prestonsburg as a base and later continue south along U.S. 23, or they can follow KY 122 or KY 7 south, bypassing Pike County, and rejoin U.S. 23/119 in Letcher County.

Garrett

Recent development has taken a toll on Garrett (pop. 150), located at the junction of KY 7 and KY 80, nine miles west of Prestonsburg. Highway improvement led to the destruction of much of the original coal town. The settlement of Garrett predates the coal boom of the early twentieth century. Originally known as Ballard, the residents of the right fork of Beaver Creek made their living through a combination of farming and lumbering. In 1910 the Elkhorn Fuel Company purchased land around Ballard, and in 1912 it began constructing a store and more than one hundred and thirty houses. Two years later, Ballard was transformed into the company town of Garrett. The post office was moved into the company store, and the town was renamed for two prominent stockholders, John and Robert Garrett.[65]

The Elkhorn Fuel Company was later joined by other coal companies operating in the area, such as Standard Elkhorn Coal and Wells Elkhorn Coal, which added to the population boom. A trunk line, the Elkhorn & Beaver Valley Railroad, connected Garrett to the Chesapeake & Ohio Railroad at Allen in 1914, and a graded highway, KY 80, reached Garrett in 1930. In order to provide for the education of the community, a number of small schools were consolidated and a central high school was constructed. By 1935 the school at Garrett was the largest in Floyd County with more than one thousand pupils.

The growth of the mining town was accompanied by the usual vices. Garrett resident Rudolph Spencer later recalled "there was usually a lot of drinking. Of course there wasn't a lot of highways then and they just laid around the streets, in the roads, and stuff and they was just drunk all the time." Such debauchery led to the rise of the Ku Klux Klan at Garrett in the 1920s. "They'd just go around with robes and burn crosses different places. And they would lay switches on peoples porches like they's going to whip them and quiet them down."[66] Union activities also brought violence to Garrett at times. Local coal operators were determined to stifle the activities of the United Mine Workers of America. In the early 1920s, when UMWA activists Sam Caddy and Sam Pasquale tried to organize miners at Garrett, their residence was dynamited.[67]

Garrett's population reached one thousand by 1940, but the 1950s brought a drop in the number of residents as mining in the area declined. In 1951 the Elkhorn Coal Company sold much of its properties in

Houses on Garrett Hill in Garrett, Kentucky.

Garrett.[68] A row of brick, frame, and stone commercial buildings, and the Garrett First Baptist Church, line the west side of the street facing the railroad tracks. Some of these are one-story native stone structures; others are brick two-story commercial buildings typical of the 1920s. Most of the original Elkhorn Coal Company houses were destroyed to improve KY 80, but twenty-three company houses still remain in the town's most interesting residential area, Garrett Hill. These two-story frame dwellings with front porches are on concrete and stone foundations and face the walkway up the hillside, the only access to these homes.

Wayland

Wayland (pop. 298) is located about five miles southeast of Garrett along KY 7 at the junction of KY 1086. Built by the Elkhorn Coal Corporation in 1912, the town was named for Sen. Clarence Wayland Watson, who was an officer of the Consolidation Coal Company and a prominent promoter of the coal industry. The center of the community was located at the confluence of Right Beaver Creek and Steele's Creek, and the commercial section was located along KY 7. The company constructed a long line of duplexes for their workers that ran parallel to Steele's Creek along the sides of KY 1086. A second residential area, consisting of single-family dwellings, was located in "the bottom" behind the commercial buildings on the opposite side of KY 7.[69]

Wayland prospered during the 1940s, but in 1954 Elkhorn Coal closed its three mines, which led to the decline of the town. Having reached a population of almost two thousand in the 1950s, Wayland had fewer than four hundred residents by 1970. The large coal preparation plant, built in 1939, was torn down in 1978. The remains are no longer visible. The tipple located at the site of the plant gave the town its mascot, the wasp. Wayland has suffered structurally from the frequent flooding of Right Beaver Creek and a declining population; however, many former residents of the community gather at the annual Labor Day homecoming festival to share their memories.[70]

Wayland High School, a large three-story red brick building with dual front entrances, was constructed as a WPA project and gained considerable fame in the 1940s and 1950s due to the exceptional abilities of its student athletes. The Wayland Wasps basketball team was considered a major power in the state from 1947 to 1964. Certainly, the pride of

Company houses, Wayland, Kentucky.

Wayland for many years was "King" Kelly Coleman. Coleman was one of eleven children born to a Wayland miner. While playing for the Wasps from 1955 until 1956 he managed to score 4,263 points and was named an All-American. After leaving Wayland, he played for Kentucky Wesleyan College and was later drafted by the New York Knicks.[71]

Like most coal towns, Wayland was a segregated community, and African Americans lived along Shop Branch, about one mile from KY 7 along KY 1086 in an area across Steele's Creek separated from the main road. It was served by a separate store, church, and school. The other main residential area of Wayland is along KY 7. Many of the original company houses remain in both residential areas of Wayland, but, unfortunately, few of the commercial buildings have survived. Opposite KY 7 in "the bottoms" are two-story wood frame homes that have been placed on elevated foundations to avoid frequent floodwaters. The current post office served as a bank from 1914 to 1931. The company store, administration building, hotel, and hospital have all been demolished.

Drift

Drift (pop. 300) is located about seven miles south of Martin, along KY 122. When the twentieth century began, a number of log cabins were located here. The trees along both the left and right forks of Beaver Creek were cut down by timber companies beginning in the late nineteenth century. Legend has it that Drift gained its name from a curve in the creek that produced numerous logjams. Coal mining operations began along the Left Fork of Beaver Creek in Floyd County during the First World War, and one of the many communities to arise during the coal boom of that era was Drift.[72]

Drift was built in 1919 and conditions were fairly primitive. During its early mining operations, the Floyd-Elkhorn Coal Company constructed small shotgun houses for miners that did not contain indoor plumbing or electricity.[73] Former Drift resident Raymond Wright recalled that as late as the 1930s he and other boys would haul water from a pump to people's homes in town. He was paid ten to fifteen cents per washtub.[74] During the 1920s, Drift was transformed into a substantial town with as many as one thousand residents. Entrepreneur W. J. "Big Bill" Turner built new, better constructed homes for miners to rent, and the Turner-Elkhorn Mining Company assumed control of Drift. In 1931 B. F. and C. D. Reed of

The Turner–Elkhorn Mining Company office building, Drift, Kentucky.

Shamokin, Pennsylvania, purchased Turner-Elkhorn and greatly expanded its operations. Unlike many coal operators, the Reeds chose to live near their mines, and B. F. Reed continued to reside in Drift until 1984. During its heyday, Drift boasted a movie theater, bowling alley, numerous stores, and a first-rate baseball park that was dedicated by major league baseball commissioner Albert B. "Happy" Chandler.[75]

There still remain a few traces of this once thriving community. The old theater, a three-story yellow brick building with art deco details, was once operated by Ernest Turner and is still intact, located on KY 122 next to the post office. The post office occupies the first floor of the W. J. Turner Building, an attached yellow brick structure built in 1949. The Drift Consolidated Grade School, built in the 1930s, is visible from KY 122 but is no longer in use. The main office of the Turner-Elkhorn Mining Company is still in use and is distinguished by its pressed metal ornamentation. It is also located on KY 122 in the center of town.

McDowell

The community of McDowell (pop. 400), located nine miles south of Martin along KY 122, predates the coal mining era. The town was probably named for Walter McDowell, a North Carolina–born schoolteacher who was instrumental in getting a post office established in 1879 when the Left Beaver region of Floyd County was relatively isolated. Coal mining began on a large scale in 1918 when the Chesapeake & Ohio Railroad extended a track line from Martin to Wheelwright. A surface road was not completed along Left Beaver until 1937.[76]

Although mining operations at McDowell were never as extensive as those at nearby Wheelwright, the community was large enough by the 1930s to sustain a high school. In 1938 a large two-story brick building was constructed to serve the educational needs of McDowell. After World War II, in an ambitious effort to provide miners with better health care, John L. Lewis of the United Mine Workers of America attempted to build hospitals in the coalfields to serve union miners. In 1956 the McDowell Miners Memorial Hospital was built, but dwindling membership forced the union to sell the hospital in 1964 to Appalachian Regional Health-care, which continues to serve the needs of people in southern Floyd County in this facility.[77]

Only a few of the original buildings remain in McDowell. Small, wooden frame miners' houses are located on both sides of Left Beaver Creek near the center of town along KY 122/680. The hospital has a modern facade, but the main portion of the original facility remains intact. Although McDowell High School was consolidated with one at Wheelwright in 1991 to form South Floyd High School, the two-story red brick building constructed in 1936 continues to be used as a grade school. Left Beaver Creek area of Floyd County is well known throughout eastern Kentucky for producing politicians. Former McDowell area resident Henry Stumbo dominated the political life of Floyd County for more than a generation. First elected as county judge in 1949, Stumbo continued in office until his death in 1978. His power and popularity allowed other members of the Stumbo clan to rise to local and statewide prominence.

Wheelwright

The largest and best-preserved coal town in Floyd County is Wheelwright (pop. 1,042), located at the junction of KY 122 and KY 305, approximately twenty-two miles south of Prestonsburg. Wheelwright spans a two-and-one-half-mile hollow along Otter Creek, settled and farmed by the Hall family in the nineteenth century. The Elkhorn Coal Company obtained the mineral rights to Hall's Hollow and began developing the area in 1916. Originally, the settlement was a tent camp and many of the Hall family worked in the early mines. Once materials were available, Elkhorn built some very basic housing and named the camp after Jere Wheelwright, president of Consolidation Coal, an affiliated company.[78]

Wheelwright was typical of the company towns found throughout the Big Sandy region in the early twentieth century. The homes built by Elkhorn were poorly constructed and lacked basic services. Housing was primitive, with floorless wood frame dwellings built from materials hauled in by the railroad. Water was obtained from hand pumps, garbage was dumped in a nearby ravine, streets were unpaved, and not all homes had electricity. Wheelwright was quite isolated and only accessible by train until the 1930s. There were no paved streets in town, and only a handful of cars made their way through the rutted, muddy roads. Miners worked a ten-hour shift and were paid seventeen cents an hour. From 1916 to 1930 Wheelwright gained a reputation as a rough community with a large male population, a mixture of races and ethnic groups and frequent violence.[79] In fact, "Bad" John Hall had served time for manslaughter before becoming the town's lone policeman in the 1920s. He reportedly killed at least seven men during his lifetime.[80]

Big changes took place in 1930 when Wheelwright was purchased by the Inland Steel Company of Chicago. The low sulfur and ash content of Wheelwright coal made it high-quality coking coal, and Inland Steel needed a regular supply for its operations at Indiana Harbor, Indiana. Inland Steel thoroughly modernized Wheelwright and turned it into a model company town, providing housing, facilities, and services far superior to those in neighboring communities. Efficiency, safety, and good working conditions were given top priority. New power substations, modern tipples, bathhouses, and showers were installed. Safety and rescue teams were organized and medical facilities built.[81]

Inland Steel catered to miners with families. The company refurbished existing houses and built new ones to high standards. A water and sewage system was constructed, streets were paved, and nearby roads were improved. Indoor plumbing was installed and new schools were constructed and remodeled by the 1940s. There was regular trash collection and incineration, fire hydrants, and new sidewalks. As more families moved to the town, Inland Steel developed community-oriented facilities. A new company headquarters and recreation center was housed in a block of buildings that were thoroughly remodeled in the style of colonial Williamsburg. Wheelwright soon had a fine library, bowling alley, inn, restaurant, and theater. All were state of the art, with modern furnishings and air-conditioning. A golf course and swimming pool were later added. By 1951 a cable television system was installed in Wheelwright, the first in the region, bringing in Ohio and Huntington, West Virginia, stations. The company invested more than $20 million in the town by the 1960s.[82] Wheelwright was touted as an ultramodern company town, featured in *Life* magazine and made the subject of a special Inland Steel publication celebrating the transformation.[83]

The population of Wheelwright changed as well. Elkhorn hired many immigrant men in the 1910s (especially Czechs, Hungarians, and Italians) to work in the mines, but their numbers declined during the 1920s.[84] Many of the miners who had worked for Elkhorn stayed on after 1930; new workers brought their families as the town was upgraded. Wheelwright became a family-oriented community with few immigrants and a large southern white and African American population. The frontierlike reputation of the town soon disappeared.

As a segregated community, Jim Crow laws dictated the layout of Wheelwright. There were separate neighborhoods and facilities constructed for whites, immigrants, and blacks. The immigrants were marginal, housed separately at first but later mixed with southern whites. The African American community was formally segregated at work and in town. Inland Steel built high-quality facilities for black miners and their families, superior to those in most mining communities, and attracted a large African American population. Hall's Hollow, a thriving black neighborhood, was developed with a school, churches, and new homes. There was a boardinghouse for black miners and four black teachers were hired for the school. African American miners worked in separate crews and had their own emergency and rescue teams.[85] Although segregated, both black and white residents of Wheelwright recall few racial conflicts. Gertrude Tyson Smith, a longtime resident, said that living in Wheelwright was "one thing I thanked God for. We didn't go by color lines. We played and enjoyed each other."[86]

By the time Inland Steel had rebuilt Wheelwright, the United Mine Workers of America had organized much of the Big Sandy region and Inland Steel's Wheelwright miners were unionized. The UMWA held regular meetings in company facilities and organized social activities, training sessions, and community work. Miners worked under the Big Sandy–Elkhorn District Agreement, and the Wheelwright Mine was a union shop. Inland Steel experienced good relations with the union and provided attractive benefits. Modernization and unionization brought many people to the Wheelwright area. By the 1950s the Wheelwright valley was full, with a waiting list for families to move into company housing.[87]

The man credited with developing the town as a model community was Emory R. "Jack" Price, the superintendent of Inland's Wheelwright operations. Price was highly respected in the community, but with a reputation as a benevolent dictator. As an Inland Steel superintendent, he ran the town. He was responsible for bringing many of the new services to the area, including electricity and water. He insisted on recreational facilities and worked hard to get the town a swimming pool, theater, and recreation center. Price also used his power to keep local residents in line. He would remind people to take care of their homes or would intervene in personal matters if he thought they would disrupt the community. Former Mayor Elmer Ferguson, who worked closely with Price, remembered him as a "man of conviction . . . If he wanted to do something, he did it." Wheelwright prospered under his leadership, and Ferguson recalled that people came "here from everywhere—New York, Chicago— just to see this place."[88]

In 1946 the U.S. Department of Interior's Coal Mines Administration created a commission headed by Rear Adm. Joel T. Boone of the U.S. Navy Medical Corps to survey health and living conditions in the bituminous coal industry. The resulting study, *A Medical Survey of the Bituminous Coal Industry* (more commonly known as the *Boone Report*) was published in 1947. The Coal Mines Administration hired documentary photographer Russell Lee and his wife, writer Jean Lee, to record life in several of the coal towns under study. Russell Lee had made his reputation as a photographer during the 1930s documenting the Great Depression for the Farm Security Administration and later recorded the events of the Second World War. The Lees visited coal miners and their families and spent time with them at work, at social events, and in their homes. One of the towns they documented was Wheelwright. Russell Lee recorded working conditions in the mines following both white and African American work teams, rescue crews, and managers. The Lees visited the homes of several miners, including Henry Armour, one of Inland Steel's best coal loaders. Armour was known as an extremely hard worker and exemplified the opportunities Wheelwright offered African Americans. In 1945 he earned more than $5,400, and when he was not mining he did home renovation, barbering, and gardening. The works they produced have left us a vivid picture of the community during its heyday when Wheelwright as a busy, friendly place, alive with work, families, and social events. The Lees' works are now housed in the National Archives in Washington.[89]

Although Inland Steel brought new life to the town in the 1930s, there was not a dramatic increase in population. The improvements in mining were typical of the highly mechanized systems that would become common by the 1950s. When Inland Steel took over the town in 1930, there were about eighteen hundred residents. Wheelwright's population peaked during World War II with about twenty-one hundred residents. More than thirteen hundred miners once worked in the operations at Wheelwright and Price, many living in nearby communities by the late 1940s. But during the 1950s things began to change, and by 1960 the population dropped back to less than fifteen hundred.[90]

Wheelwright continued to produce coal, but production slowed. In 1965 the era came to an end. Inland Steel moved its coal operations to Cicero, Illinois, and sold its Kentucky properties to the Island Creek Coal Corporation. Wheelwright ceased to be a company town. Rents were raised dramatically from two to ten dollars per room per month. Island Creek took little interest in housing or facilities and soon sold the town. As Wheelwright passed from company to company in the 1970s, mining profits decreased, population declined, and the town quickly lost its reputation as a model community. Mountain Investment Incorporated assumed the properties at Wheelwright for $1.3 million and sold the town to the Kentucky Housing Corporation in 1979. Houses were then sold to individual residents, facilities were leased, and services completely disappeared. By the 1970s the mines at Wheelwright began to close, although the operations at Price continued until 1991. Many people continued to live in Wheelwright and work in nearby mines, but the loss of steady employment drove others away.[91]

Wheelwright stills has the feel of a thriving company town. It has maintained a small but loyal population over the years, and many houses and buildings are still in use. At the entrance to the town off KY 122 there is a post office and a native stone building that was once the movie theater. The raised bank of land on the left side of the road is the old railroad grade where coal and passenger trains made their way through the town. One-quarter mile farther are three red brick two-story school buildings constructed during the late 1930s as WPA projects. The gymnasium building is still in use, but the other two have recently been closed. Farther ahead is the swimming pool, reopened in the 1990s after years of neglect.

The road winds through the town, with the railroad grade on the left and houses on the right. The homes vary in size and design, some were built in the 1920s and others after Inland Steel took over. Most are one-story, three- or four-room frame cottages on post foundations with front porches. They were renovated in the 1930s when Inland Steel acquired the town, adding water, electricity, and gas. Some had telephones as well. The homes in Wheelwright are a bit varied; some were remodeled in the 1940s and others have been modified since the company left town. The typical miner and his family paid about $11.50 a month for rent. Buyers paid about $1,200 for these homes when they were sold in the 1970s.

Just before the center of town, behind the discount store, a road leads up the left to Hall's Hollow, where there is still a small African American community and a lively Baptist church. At the entrance to the hollow is a boardinghouse with a large porch where single and newly arrived African American workers resided. The one-story brick building at the head of the hollow was once Lee Hall's General Merchandise Store. The long front porch, with its brick posts and steps, was a popular gathering place for miners and their families. On the main road, opposite the entrance to Hall's Hollow, is a small, one-story native stone building constructed by the WPA as the colored school during the 1930s for Wheelwright's African American children. Today it is a private residence.

Between Hall's Hollow and Maple Street is the central commercial district of Wheelwright. Many of the buildings along Main Street date from the 1920s but were remodeled and brick-veneered by Inland Steel in 1941. Many are of the colonial revival style, a style that became popular in the early twentieth century. A national interest in historic preservation during the 1930s influenced the choice. Wheelwright's colonial revival buildings reflect the traditions of colonial Williamsburg. They are the largest and most imposing buildings in the town but are in poor condition today.

There is a veteran's monument on a small lot where the Inland Steel central offices were located. The large complex next to the monument was the company clubhouse. It is a two-and-one-half story structure with three sections; four white columns and a recessed front entrance separate

the two side units. Once an elegant and lively place with a fine restaurant, hotel, lobby with small shops, and bowling lane in the basement, the structure that once housed Wheelwright's library is now abandoned, with piles of books scattered on its floors. In 1980, however, it was listed on the National Register of Historic Places.

Directly across the street is the community building, a massive structure built in 1916 by Elkhorn Coal and remodeled by Inland Steel in the 1940s. It has a two-story portico extending the full length of the facade, supported by a series of ten fluted columns. This was one of the most important buildings in town and once housed a post office, barbershop, restaurant, and soda fountain. The second floor was a large dance hall where social events and community meetings took place. This site was also listed on the National Register of Historic Places in 1980.

Opposite page, top: Coal miner Henry Armour and his family, Wheelwright, Kentucky, 1946. Russell Lee Collection, National Archives Still Pictures Branch. Opposite page, bottom: Miners during shift change at the Inland Steel mines, Wheelwright, Kentucky, 1946. Russell Lee Collection, National Archives Still Pictures Branch. Above: Remains of Inland Steel buildings at Wheelwright, Kentucky.

Across from the clubhouse, on an elevated site, is the Wheelwright Methodist Church, originally built as a nondenominational community church. Steps lead up the hillside to the front entrance, which is dominated by a large bell tower. The one-story church with its arched windows and stained glass played an important part in the social life of Wheelwright residents. The bells chimed every evening at six o'clock and an amplifier broadcast organ music. This provided a pleasant background for people as they gathered downtown, visited with neighbors, or frequented the community center.

Next to the church is a house that was built for the superintendent and general manager. It also reflects some of the colonial revival style and is the only major house made of brick in Wheelwright. Beyond the commercial district is another residential area that winds through the hollow. At the top of the incline is a small one-and-one-half-story wood frame store building with a false front facade.

The Wheelwright mines are located on the hillside opposite Hall's Hollow. A road alongside the Methodist Church leads to the site, where an Inland Steel sign marks the entrance. From the mine site, there is a terrific view of the town and surrounding countryside. Directly across is a view of the downtown commercial district and the winding road that leads to Hall's Hollow.

In 1940 the neighboring town of Burton was added to Inland's operations and similar improvements were made. Production increased during the Second World War and Inland Steel continued to expand. In 1949 an entirely new mining operation was begun at Price (about four miles north on KY 122). The site was named for Jack Price, the superintendent of Inland's Wheelwright operations.

Inland Steel expanded its mining operations beyond Wheelwright in the 1940s. Price was established in 1923 but was taken over by Inland Steel in 1948 when a new mine site was developed. The tram, tipple, and processing center were completed by 1951. This was the last major investment that Inland Steel put into the Wheelwright area. The site was ultramodern with state-of-the-art facilities and equipment. A chute and conveyor system brought coal down from the hillside mines, and a large coal preparation plant processed the coal. There was a four-track loading area and storage yard for coal cars. The Price site became the largest mine on the Chesapeake & Ohio line, loading more than three thousand fifty-ton hoppers a month at its peak in the 1950s. Price was often featured in the railroad's advertisements and promotional literature because of its modern, clean appearance.[92] A good overall view of the area can be seen by following the left turnoff over the small bridge and continuing about halfway up the mountain.

Wheelwright's heritage has been kept alive through the efforts of many of its present and former residents. Before Wheelwright High School closed in 1996 it was the home of a student publication called *The Mantrip*. For ten years, students interviewed people in the community and recorded their memories in this publication. It had subscribers all over the United States and Canada.[93] In 1996 *Lexington Herald-Leader* reporter Michelle Patterson-Thomas located some of the people the Lees had photographed and interviewed in the 1940s. Like many, they had fond memories of growing up and living in the town but eventually had to leave. The family of Henry Armour took advantage of the new opportunities available to African Americans in the 1970s. The children moved away to pursue college degrees and now live in Akron and Indianapolis. They credit their father with providing the work ethic and motivation that allowed them to succeed.[94]

Although Wheelwright was established during the great coal boom of the 1910s, it is a fine example of a modern mining community from the last phase of company coal town life in southern Appalachia. Wheelwright has recently undergone a bit of a resurgence with the opening in 1993 of the Otter Creek Correctional Facility, a minimum security prison. The facility has provided new employment opportunities for area residents but may have an impact on some of the remaining mining structures.

U.S. 23 up the Levisa Fork

From Prestonsburg south to Pikeville, U.S. 23 follows the Levisa Fork and the CSX tracks. Pike County is the largest county in the state, and coal has dominated its economy for much of the past one hundred years. Pike County (pop. 68,736) was formed in 1821 and named for Zebulon Pike, who explored the American West in the early nineteenth century. The county was primarily agricultural for the next fifty years, but by the 1870s a timber industry developed and steamboat trade connecting Pikeville with Catlettsburg brought further commercial expansion. This transition brought economic and political rivalries that sometimes became violent, and in the 1880s Pike County was the site of the famous

Hatfield-McCoy feud. An early-twentieth-century coal boom came as the Chesapeake & Ohio Railroad opened the Marrowbone coalfields in 1904. A few years later the eastern section of the county along the Tug Fork was penetrated by spurs from the West Virginia line of the Norfolk & Western Railway. Although the mid-twentieth century depressed the economy, there was a great resurgence during the 1970s as a second coal boom brought jobs, people, and investment back to the county. As a result Pikeville, the county seat, became a leading financial and commercial center. The boom was short-lived, but hundreds of small mines and several large strip mines presently operate in the county. In recent years Pike County has produced more than thirty-five million tons of coal annually, 20 percent of the state's total. Productivity is high, but employment opportunities are limited. During the 1990s, Pike County has experimented with light industry, retail expansion, and tourism. Visitors will find good accommodations and services in Pikeville, as well as a variety of scenic drives through remote areas of the county.[95]

Pikeville

Pikeville (pop. 6,300) is located south of the junction of U.S. 23 and U.S. 119, along the Levisa Fork. It is a regional commercial center and the seat of Pike County, an area central in the development of the Big Sandy Valley coal industry. It is a good location from which to explore the greater Big Sandy region and has several interesting historic sites. Settled in 1823, Pikeville (Piketon) was a small, quiet river town serving local farmers. When James Garfield and eight hundred Union troops camped in Piketon during the Civil War, the town had just over one hundred residents.[96] In the 1880s Piketon became the city of Pikeville as riverboat trade increased and a timber industry developed on the Levisa Fork. Improvements followed and regular steamboat service began between Pikeville and Catlettsburg, a distance of 120 miles. In 1905 the Chesapeake & Ohio Railroad reached Pikeville, replacing the river as the main means of transportation. With the growth of the coal industry, Pikeville's population began to increase, from about five hundred residents in 1900 to more than two thousand in 1920. The town was often noted as a scenic site with "long narrow streets, surrounded by thickly timbered countryside that ranges from the hilly to the mountainous; [and] neighboring roads [that] reveal scenes of wild, almost breathtaking beauty."[97]

The automobile age brought further growth. Two federal highways, U.S. 23 and U.S. 119, were completed in the 1930s linking Pikeville with Paintsville, Jenkins, and Williamson. Under the New Deal, the WPA helped fund electric lights for Main Street, new public buildings, and the restoration of the county courthouse. The population of Pikeville doubled once again and held steady during the 1940s.[98] It remained an important town but began to suffer as coal prices dropped in the early 1950s and employment in the coal industry declined. To make matters worse, the area suffered two of the worst floods in recorded history in 1957 and 1963. The downtown commercial district was submerged and hundreds of homes were badly damaged.[99]

The Great Society programs of the 1960s brought renewed attention to the area. Pikeville College attracted reformers who worked with Volunteers in Service to America (VISTA) and local agencies to develop citizen participation in government. Conflicts between local leaders and outside reformers created tensions, even leading to the arrest of several under Kentucky sedition laws. But the coal boom of the 1970s changed Pikeville. A spurt of development brought new stores, facilities, and suburban-style housing. Pikeville became a financial hub. As local bank assets reached $300 million in the mid-1970s, the city gained a reputation for having the highest per capita number of millionaires of any city in the United States.[100] Many people still consider Pikeville a distinctly different city, wealthier and more cosmopolitan than the surrounding communities.

In order to prevent future floods, a major engineering project known locally as the "cut-thru" began diverting the Levisa Fork of the Big Sandy River west of town in the 1970s. This diverted not only the river but also railroads and highways, creating new flatland for urban development. The project took fourteen years to complete as twenty-three million cubic yards of rock were blasted away at a cost of more than $60 million. It created a channel 523 feet deep and 1,300 feet wide and opened up 390 acres of usable land in the city. A historical marker notes that this was the largest earth-removal project ever undertaken in the United States and is "second in the world only to the Panama Canal." Today, visitors can view the cut-thru from a scenic overlook near Bob Amos Park.

Pikeville was the center of the famous feud between the Hatfields and the McCoys. Both families made their living in the Big Sandy River region by farming and logging. Though the Hatfields and the McCoys lived on both sides of the Tug Fork, the Hatfields tended to live in West Virginia while the McCoys were more numerous in Pike County—especially in the vicinity of Blackberry Creek.[101] The feud developed in 1869 when William Anderson "Devil Anse" Hatfield formed a company so that he could enter the lucrative timber trade along the Tug Fork. In preparing his business, he initiated a lawsuit claiming land owned by the Cline family. His success in the case against the Clines helped lay the groundwork for the feud.[102]

Problems between the Hatfields and McCoys went as far back as the Civil War, but the feud actually started in the 1880s. There had been a series of minor incidents, such as a dispute over a pig and the failed romance between Johnse Hatfield and Rose Anna McCoy, but none had resulted in any extensive fighting. This all changed in August 1882 when three sons of Randolph McCoy brutally attacked Ellison Hatfield, the brother of Devil Anse. When Ellison died two days later, Devil Anse called for revenge. The McCoys were hunted down, tied to papaw trees on the Kentucky side of the Tug Fork, and executed. The bodies were riddled with bullets. A Kentucky grand jury indicted twenty Hatfield partisans who were thought to have taken part in the attack, but there was no serious attempt to apprehend them.

Things were quiet for a while, but five years later, during the election of 1887, the flames of the feud were stirred into an inferno. By all accounts it was attorney and political activist Perry Cline who rekindled the smoldering embers. Years earlier Cline had lost thousands of acres of timberland to Devil Anse, and Cline's sister Patty had married Asa Harmon McCoy. Cline moved to Pikeville, where he took up the practice of law.[103] He worked hard to win votes for Kentucky gubernatorial candidate Simon B. Buckner, and when Buckner won the election, Cline pushed him vigorously to pursue the extradition of the Hatfields from West Virginia and offered a sizable reward for their apprehension.[104] By the end of the year, Pike County posses were crossing into West Virginia, seizing Hatfield partisans and returning them to Kentucky to stand trial.

In order to prevent the convictions of friends and family members, Devil Anse Hatfield organized a raid on the home of Randolph "Old Ranel" McCoy. Launched on New Year's Day 1888, the attack led to the death of two of Randolph's children, the brutal beating of his wife, and the complete destruction of the McCoy homestead on Blackberry Creek. The New Year's raid attracted the attention of the press throughout the region and greatly inflamed passions on both sides of the Tug Fork. Kentucky posses intensified their efforts, and small-scale battles were fought between partisans for the next year and a half. Though West Virginia authorities challenged Kentucky's right to try the Hatfields, the Mountain State was rebuked by the U.S. Supreme Court. Eight Hatfield partisans were eventually sentenced to life imprisonment and one was executed.[105]

The feud drew national attention and became an important part of the history and folklore of the Big Sandy region. Newspaper reporters described the Tug Valley as an isolated, savage part of the country, sometimes comparing it to the wilds of Africa: a lawless area without basic social institutions. Many stereotypes about the people of Appalachia stem from these accounts. "Hillbillies" were portrayed as violent, vengeful, primitive people who lived in untamed wilderness. The history of the feud was soon forgotten, but it made its way into folklore and legend. Books, radio, and film productions kept the feud alive. It entered our popular culture with images of barefoot bearded gun-toting moonshiners, shotgun weddings, and family grudges that always led to violence.[106]

Today descendants of the Hatfields and McCoys still retell the story of the feud and argue over the details. During the last years of the twentieth century, many of the sites of the feud have been listed on the National Register of Historic Places, and both Kentucky and West Virginia historical markers have been set at some of the locations. Pikeville, Kentucky, and Matewan, West Virginia, have developed the feud as a tourist attraction with driving tours, reenactments, reunions, and celebrations. Pikeville's Hillbilly Days, held in mid-April every year, brings people together from all over the world to celebrate hillbilly culture, and stereotypical images of the feud can be seen everywhere.

Some of the sites lie within Pikeville in a newly designated Hatfield-McCoy Historic District. At the Dils Cemetery on Chloe Road, a Kentucky highway marker tells the story of the feud. This is the site of more than five hundred graves, including those of clan leader Randolph McCoy, his wife and children, and members of the Dils family. The Pike County courthouse and jail was the scene of the Hatfield clan trials for the murders of Tolbert, Randolph Jr., Pharmer, and Calvin McCoy. This three-story native-stone structure was built in 1889 and has undergone several expansions and renovations that have added a red tile roof and modern window trim. Outside, in the courthouse square, a statue commemorates the soldiers who served in the First World War.[107] The defendants were kept in the jail during the trial and all received life sentences except for Ellison Mounts, who was hanged on 18 February 1890. The hanging took place near today's Kentucky Avenue on the campus of Pikeville College. A Kentucky highway marker indicates the site of the execution.[108]

Markers have also been placed at several spots in the Blackberry Creek area of Pike County. They include the site of Randolph McCoy's house, the papaw tree incident, the hog trial, and the Election Day fight. They can be located with the help of a driving tour brochure available from the Pikeville–Pike County Tourism Commission or the Matewan Development Center. Throughout the Tug Fork valley there are numerous unmarked sites where the homes of feud participants once stood or where incidents in the feud took place.[109]

There are a few other places of interest in Pikeville that can be seen quickly on a brief tour of the city. The main route into town is Hambley Boulevard (Business U.S. 23/460), accessible from the downtown Pikeville exit of U.S. 23. At the city park on Huffman Avenue, the Tourism Commission can provide a county roadmap and a city walking tour guide. The park is located within the Huffman Avenue Historic District, where James A. Garfield's Union troops camped during the Civil War. It includes the First Presbyterian and the Pikeville United Methodist Churches, both listed on the National Register of Historic Places. The nearby College Street Historic District, placed on the National Register of Historic Places in 1984, has a number of residences built in the American foursquare tradition.

Hambley Boulevard follows the former route of the Chesapeake & Ohio Railroad through the town. At the intersection of Hambley and Division Street is the old C & O Railroad depot and passenger terminal, also listed on the National Register of Historic Places. The two white-trimmed classical revival brick buildings faced the tracks and were once connected by a long shed roof, supported by wooden columns that shaded the platform. Luggage and freight filled the wooden carts along the platform. Passengers arriving in Pikeville could board buses behind the depot to take them through the city or to nearby towns. The two buildings now house several city government offices. From Hambley Boulevard, Elm Street leads up the hill to Pikeville College, founded in 1887 to serve the city's Presbyterian community (it also houses a small collection of local history materials). Halfway up the hill overlooking the city is the Augusta Dils York Mansion, a two-story cream brick residence with dark mortar work, built in 1918. It is an example of classical revival architecture and features a clay tile mansard roof and wraparound porch.

The Huffman Avenue Historic District, Pikeville, Kentucky.

This Victorian-style home on a prominent hillside location has become a well-known local landmark visible from around the city and is also commonly believed to be haunted. John R. Dils settled in Pikeville in the 1840s, established a dry goods business, and invested heavily in coal and timber land, buying some for as little as two and a half cents an acre. He was rumored to be the wealthiest man in Pike County in the late nineteenth century.[110] His daughter Augusta Dils married attorney James M. York, who helped defend Randolph McCoy during the feud. Today, the mansion is opened seasonally to visitors. For a glimpse of what Pikeville looked like before the cut-thru, head north of the city on Bypass Road, where there is a small native stone building constructed by the WPA in 1937 as a school for Pikeville's black community.[111] Today it houses a community center named for Perry Cline, a former school superintendent. A small African American community still lives near the wooden footbridge that crosses what remains of the Levisa Fork and connects the area with the Scott Avenue Historic District, a collection of bungalow and foursquare residences from the 1920s.

Pikeville's Main Street is typical of the region's commercial centers, with a variety of buildings dating from around the turn of the twentieth century, and was placed on the National Register of Historic Places in 1984. The site of the new civic center was where riverboats once docked. The site of the new civic center at the end of Main Street is the Italianate-style York house, one of the oldest residences in Pikeville, dating from the period before the Civil War. Pikeville continues to be a thriving city, one of the few in the region to grow and prosper in recent years. Because of its long and varied history, it has had several historic districts listed on the National Register of Historic Places that showcase different periods of development. Pikeville's annual Hillbilly Days celebration each April draws as many as 100,000 people to the city.[112]

Just south of Pikeville is the CSX Railroad's Shelby Yard, an important marshaling yard for area coal mines. It is located at the junction of Old U.S. 23 and KY 122 across a concrete bridge. The Shelby Yard was once the site of several water tanks, a coaling station, a three-bay round-house and turntable, and company offices. As mining expanded during

Members of the McCoy family are buried in the Dils Cemetery in Pikeville, Kentucky.

the Second World War the yard was enlarged and modernized. Most of these structures have been demolished, and today the Shelby Yard is a fifteen-track yard with offices and a single-stall engine house. There is still a considerable amount of traffic through the yard bringing in coal from Pike County and Southwest Virginia mines.[113]

Many of the area's coal towns can be reached from Pikeville. The city offers travelers the options of heading south toward Virginia, east into West Virginia, or continuing west along U.S. 23/119 toward Letcher County. South of Pikeville U.S. 460/80 follows the Russell Fork of the Big Sandy River and heads toward the Virginia border. About nine miles south of Pikeville, KY 195 can be taken south through the Marrowbone coalfields to the county's best-known coal town, Hellier. KY 195 connects many of the towns that were along the Chesapeake & Ohio's Marrowbone branch, which opened in 1908. The Marrowbone area was booming in the 1910s and 1920s, but today it is an isolated, quiet part of Pike County. A few hundred residents inhabit the once bustling mining camps of Wolfpit, Ratliff, Dry Fork, and Rockhouse. Following U.S. 119 toward West Virginia, travelers can visit the county's best preserved coal town, Stone.

KY 195 to Hellier

From Pikeville, KY 195 passes through several small coal towns: Wolfpit, Ratliff, and Rockhouse. The most prominent building in the area is the L. E. Ratliff General Merchandise Store, located at the entrance to Lookout alongside the railroad grade. It dates from 1901 and for decades served as a railroad depot, post office, and general merchandise store. Until the mid-1950s, steam engines made two daily trips to Pikeville, returning with mail, passengers, and commercial goods. The center portion of this one-story brick building appears to be the original store, later expanded with two side additions. Its unique tin-plated exterior and false-front facade dominate the entrance to Lookout. The store had two large front display windows where merchandise was displayed and two entrances—one to the store, the other to the post office. The store closed in 1973, but the building still houses the local post office. Just past Lookout is Henry Clay. The Henry Clay Coal and Coke Company developed both of these towns while leasing land from the Big Sandy Company after 1907.[114] There are about forty houses still occupied in the area.

The best known of Pike County's coal towns is Hellier (pop. 300), located at the mouth of the Brushy Branch of Marrowbone Creek, about twelve miles south of Pikeville on KY 195. Hellier was one of the earliest company towns built in the county. It grew as the Marrowbone coalfields were developed in the 1910s and later served as a commercial center for many of the nearby mining towns. In the 1920s Hellier was a booming place and gained a reputation for its frontierlike atmosphere and violence.

Charles Edward Hellier was from a prominent Boston family, well educated and well traveled. He graduated from Yale, continued his studies at the University of Berlin, and later received a law degree from Boston College. Hellier was connected with the Warren Delano family, a major investor in the Elkhorn coalfields, and together they acquired seventy thousand acres of land in eastern Kentucky in the 1890s. In 1902 they formed the Big Sandy Company (with Hellier as president) to develop their land in Pike County. They convinced the Chesapeake & Ohio Railroad to build an extension of track from Whitehouse in Johnson County

This store and post office in Lookout, Kentucky, served area mining town residents for most of the twentieth century.

COAL TOWNS IN EAST KENTUCKY

south to Ashcamp on Marrowbone Creek, a distance of more than one hundred miles. From 1904 to 1906 they constructed Hellier, a planned company coal town named for Ralph Augustus Hellier, brother of Charles and general manager of Elkhorn Coal. Ralph came to Pike County in 1894 and helped develop some of the earliest mines along Marrowbone Creek. The coal along Marrowbone soon became known for its high-quality coking properties.[115]

Hellier grew rapidly, especially during World War I. There were three main mines near Hellier and long rows of coke ovens. Other mines opened as the Big Sandy Company leased land to the Edgewater Coal and Coke Company, the Greenough Coal and Coke Company, and the Pike Coal and Coke Company. Employment in these mines drew more than seven hundred people to the town by 1910, and the development of nearby sites made Hellier a thriving community of more than two thousand people by 1930.[116]

Elkhorn Coal recruited locals to work the mines but soon brought immigrant labor to meet labor shortages. Hellier had Hungarian, Polish, and Italian workers as well as a small African American community. The early residents were young, single men, and Hellier became known as one of the roughest places in the county. Stories of shootouts, murders, and feuds became legendary. Much of the violent reputation of Hellier was the result of a murder in one of the town's restaurants in the 1920s. Several violent incidents followed as retaliation for the killing. In the 1930s two Civilian Conservation Corp projects brought more young men into the Hellier area. There were occasional fights between the CCC boys and locals, which added to the town's negative image. Because of these incidents, many Pikeville residents were reluctant to travel in the Marrowbone area and were often fearful of its residents.[117]

The small wooden railroad depot that once stood at Hellier was a busy place as businesses grew to serve the many coal camps along Marrowbone Creek. By the 1920s it took the entire day to unload freight and transfer it to local establishments. The arrival of these shipments was an event that attracted large crowds. Along with mining equipment and mail there were deliveries of foodstuffs, medicines, furnishings, and luxury items. Children gathered at the depot just to watch the variety of goods being delivered to local merchants. The entire Marrowbone area was isolated from the rest of the county and had to be self-sufficient. Hellier had a bank, small hospital, three dry-goods stores, barbershop, jail, police force, pool hall, schools, restaurants, and a movie theater. After the repeal of prohibition, Hellier had restaurants serving alcoholic beverages and playing phonograph records for dancing. Several automobiles were shipped by train and served as taxis in the Marrowbone area. The creek beds were used as roads, and only the most adventurous drivers would attempt the journey to Pikeville, a trip that took most of the day. It was not until the 1940s, with the completion of several graded roads in the county, that Hellier became accessible.[118]

Although Hellier grew and prospered, it also suffered from several natural disasters. A major flood struck the Marrowbone area in 1926, drowning seven people at Edgewater and washing their bodies miles downstream. A second flood in 1928 caused extensive damage to homes and businesses in Hellier. Fire was also a problem, as most of the local structures were made of wood. In the 1930s one fire destroyed fourteen homes and businesses in the town.[119] Residents stored barrels of water to fight small fires, but Hellier did not have a fire truck until the 1940s.

Hellier boomed for twenty-five years, but the Great Depression had a dramatic impact. Many of the mines at the nearby coal towns of Henry Clay, Edgewater, Alleghany, and Lookout closed. A Pike County merchant, G. C. Ratliff, noted in a 30 April 1930 diary entry that it was "distressing to go through the coal camps and see the empty houses and the ragged, half-starved families that inhabit the occupied ones." A later entry noted the "hundreds of miners, almost on starvation, . . . gathered at commissaries, coal company offices, etc., eager to hear a report that the mines would start."[120] There was little work available, Hellier's businesses suffered, and the population declined. One of the few other sources of income was the gathering and drying of ginseng roots, which were used in medicinal potions. With the loss of work, the town lost more than one thousand residents.

By 1950 there were only a few hundred people living in Hellier. The remaining mines were sporadically worked but fell victim to declining coal prices. Many of the abandoned and unoccupied homes were destroyed. In 1960 the Blue Diamond Coal Company closed the remaining mines and demolished the last of the coke ovens. The mineral rights were sold to Bethlehem Steel, who began extracting the remaining coal from the other side of the mountains. In 1963, *New York Times* reporter Homer Bigart, who toured the region, wrote that the Marrowbone Creek area was "a string of ghost towns," and the "Hellier city hall is abandoned and only a few stores remain." Since the 1960s, the town has maintained a population of a few hundred residents.[121]

Most of Hellier has been lost, but a few interesting buildings remain. In the center of Hellier on KY 195 there is a two-story painted brick and hollow block building that was once the Hellier Service Station and Barber Shop, constructed in 1924 and operated until the 1990s. Across the street is a one-story brick building with two entrances and large display windows. It was the location of many small stores over the years, and the fading painted lettering from the Fair Store and Granny Re's Grocery are still evident. This was the commercial center of Hellier. Up the hillside is a small group of houses and the remains of the Hellier School, a brick two-story structure completed in the 1930s as a WPA project. A scattering of about forty homes line the road behind the store and alongside the school, but most have long since disappeared.

High on the hillside, directly across from the Missionary Baptist Church, is the old Hellier cemetery, which is now on private property and overgrown with grass and weeds. Most of the graves date from the late 1910s and early 1920s, but many of the tombstones have been overturned

This grocery store was located in the commercial heart of Hellier, Kentucky.

and broken. On KY 195, in the wooded area along the railroad tracks to the north of town, are the ruins of a tipple. Little else remains of Hellier, but KY 195 parallels Marrowbone Creek and there are a series of former coal camps in the immediate area, including Alleghany and Edgewater.[122] Each have a few buildings remaining. The former coal town of Alleghany is located along the Castle Fork of Marrowbone Creek about one mile above Hellier. The town was developed in 1905 by the Beddow Mines Company, which later became the Pike Coal and Coke Company. In 1912 the Alleghany Coke Company acquired the town, and a post office and railroad depot were named Alleghany. There are about forty closely spaced wood frame houses dating from the 1910s.[123]

Edgewater

The Edgewater Coal and Coke Company developed mines on land leased from the Big Sandy Company in 1907 one-half mile southeast of Hellier, at the present site of Edgewater (pop. 100). In 1911 the Elkhorn Consolidated Coal and Coke Company, under the direction of Fon Rogers and his brothers, built coke ovens and improved the town. At the time there were twenty houses, a store, and about sixty coal hoppers in the town. The Rogers Brothers Coal Company expanded the site and later developed mines at Virgie and Burdine.[124] By 1920 there were about one hundred new homes in the area. Today little exists of Edgewater. To reach the site of the former town, follow Edgewater Road, which is at the south end of Hellier just past the Missionary Baptist Church. After about one-half mile there are a few building foundations on the right side of the road.

KY 80 to Elkhorn City

Travelers can reach Elkhorn City from Pikeville by continuing south on U.S. 460/80 or by following KY 195 from Hellier to KY 197 and heading east. Elkhorn City (pop. 1,060) lies on the Russell Fork of the Big Sandy River at the mouth of Elkhorn Creek, south of Pikeville. The town developed after the Chesapeake & Ohio Railroad arrived at the site in 1907. Elkhorn City was the southernmost point on the line, and the railroad's coal marshaling yard sorted and stored coal from the many surrounding mines. The main line of the Clinchfield Railroad later joined the Chesapeake & Ohio here.

Elkhorn City prospered as long as the mines boomed but suffered by midcentury. The downtown commercial district's streets are lined with one- and two-story brick buildings typical of those found through-out the region in the early twentieth century. There is still some coal hauling traffic through Elkhorn City, but activity at the rail yard has been greatly reduced in recent years. The Elkhorn Yard had a modest engine servicing facility and several truck-dump coal loaders set along-side the tracks. The Elkhorn City Railroad Museum, housed in a small concrete block building at 100 Pine Street, has displays portraying the area's railroad heritage.[125]

Elkhorn City is also the gateway to the Breaks Interstate Park, located on the Kentucky-Virginia border. The Pine Mountain River Can-yon of the Russell Fork is the largest river canyon east of the Mississippi and features a five-mile cut where the river descends 350 feet over boul-ders, creating rapids and small waterfalls. There are 1,000-foot palisades covered with rhododendron and pine on either side of the canyon. Two miles southeast of the Kentucky state line is a sandstone formation known as the Towers that rises 1,600 feet above the river. According to Shawnee legends, there is a great cave in the breaks that was used for protection during battles with the Cherokee. Jonathan Swift's legendary silver mines are also supposedly located in the breaks area. Rail fans will enjoy watching coal trains make their way over the two bridges and through the four tunnels visible nine hundred feet below from the many observation points in the park. The park is open year round, has lodg-ing and restaurant facilities, and features 4,600 acres of breathtaking scenery.[126]

The Tug Valley and Stone, Kentucky, on U.S. 119

From Pikeville, U.S. 119 can be followed east into West Virginia. Just before reaching Belfry, KY 199 cuts south, following a Norfolk-Southern spur to a series of coal towns along Pond Creek. Stone is located on KY 199 about one mile south of U.S. 119. The eastern part of Pike County lies in the Tug Valley and historically has been tied to development in West Virginia. As the Norfolk & Western Railway made its way through the Tug Valley in the last years of the nineteenth century, spurs from the main rail line reached across the river into Kentucky and opened areas of eastern Pike County to mining. As the area along Pond Creek was developed, Stone became an important company coal town. The use of state-of-the-art technology, electrification, and brick (rather than stone) construction gave Stone a more modern appearance than other communities built in the 1910s. Stone is Pike County's best-preserved coal town.

Stone

Established in 1912, Stone (pop. 300) was named for Galen L. Stone of Brookline, Massachusetts. Stone was a partner in a major Boston investment house, Hayden, Stone and Company, and chair of the Pond Creek Coal Company. The Pond Creek Coal Company opened mines along Pond Creek, which flows into the Tug Fork. The company acquired more than twenty-two thousand acres of land in east Kentucky, and in 1912 it built the coal company town of Stone.[127] Most of the buildings in Stone were constructed in the late 1910s, but many houses were added as the town expanded. By 1920 the company had opened eight mines in the area and hundreds of people had moved into the town. Stone continued to grow, boasting a thousand residents by 1930, and it became an important center for the many mining communities along Pond Creek. The railroad tracks wound through the hollow, connecting the mines with the main Norfolk & Western. Alongside the tracks, rows of company housing, a commercial center, mine sites, and facilities were constructed. The commercial district contained a company store, YMCA, and company offices.[128] Managers and office personnel lived north of the commercial district, while common miners and their families lived to the south. Houses for managers and office personnel were built on the club lawn, where the railroad and the creek were the farthest apart. These houses were bigger, made of better materials, and were placed on larger lots than the common miners' homes. Most are two-story dwellings, built in the American foursquare tradition. There is little variation among these homes, with the exception of some roofing details and porch enclosures. One large bungalow in the north served as a clubhouse. Most of the miners lived south of the commercial district along the railroad tracks that ran through the town. Their homes were all built around the same time and share the same basic design: two-story frame and weatherboard with front porches facing the railroad tracks. Because they were owned and maintained by the company, there was little variation, and few changes or modifications have been made over the years.

Because of its isolation Stone was self-sufficient. It had a powerhouse, modern concrete tipple, machine shops, and warehouse. Pond Creek Coal's heavy technological investment in Stone suggests that it hoped to mine the area for many years. The company used the latest electric locomotives and mine cars, rather than just men and mules, to retrieve coal. The tipple was made of concrete at a time when wood was still commonly used. Stone also had a 200,000 gallon water tank, oil storage buildings, and an ice plant.

In 1922 the town and nearby mines were acquired by the Fordson Coal Company, a subsidiary of the Ford Motor Company. Henry Ford was frustrated by coal strikes disrupting automobile production and decided to gain control over many of the natural resources needed for his River Rouge complex in Detroit, Michigan. Ford acquired rubber plantations in Brazil, iron mines in the American West, and coal mines in Kentucky and West Virginia.

A few years earlier, Galen Stone had made the acquaintance of Joseph P. Kennedy and had invited him to come to work at Hayden, Stone and Company. Stone leaked vital information about the pending deal with Henry Ford to Kennedy and helped him borrow money to invest heavily in Pond Creek stock. Kennedy bought fifteen thousand shares at $16 a share in mid-1922. He bought the stock on margin, meaning that he only had to put up $24,000 and wait for the price of the stock to rise. In December the *New York Times* carried a story suggesting that Ford was interested in Pond Creek Coal, and the price of the stock began to jump. When Kennedy sold all of his shares at a price of $45 a share, he realized a $650,000 profit on his original investment of $24,000. This put Kennedy well on the way to becoming a very wealthy man. As historian Doris Kearns Goodwin noted in her 1987 work, *The Fitzgeralds and the Kennedys,* Joe Kennedy had amassed nearly "three quarters of a million dollars, more than his father and grandfather had accumulated in the course of their two lifetimes." Stone soon retired from the coal and investment business, and Kennedy established his own business with his new fortune.[129]

Meanwhile, with the purchase of Pond Creek Coal, Ford had acquired more than six hundred company houses in the eight-mile stretch from Hardy to McVeigh. Ford managers from Detroit surveyed the properties and began to make improvements immediately. A crew was hired to paint all the houses in one uniform color scheme: slate gray on the outside and off-white on the inside. In the central commercial district the company built a red brick theater, pool room, dry goods store, and restaurant. The company provided regular services to the residents of Stone: lawns were trimmed, flowers and plants were cared for, and fences maintained. Ford raised the base pay to five dollars a day, which was well above the three dollars that prevailed in the area. And, of course, Ford encouraged the purchase of new Ford automobiles among his employees. In Stone the company added common garages to the homes of managers for their new automobiles. To provide more social activities in this isolated area, the company organized a band, sports teams, and an employee clubhouse. Longtime Stone resident Radar Hale remembered the Ford era as the best years for the people living along Pond Creek. Great improvements were made and hard work and efficiency were well rewarded. Hale recalled that Henry Ford himself visited the town in his private rail car. It was said that as the train approached Stone, Ford stood on the coach's platform and threw silver dollars to the crowds. During his visits to Stone, Ford stayed at the clubhouse, the general manager's house, or a nearby farm.[130]

During the great manufacturing boom of the 1920s, Fordson Coal guaranteed Ford a vital resource needed to manufacture automobiles, so Stone prospered. After the stock market crash of 1929, production at Ford slowed and occasionally halted. By the early 1930s the Ford Motor Company underwent financial restructuring and began to sell its many subsidiary producers. In 1936 Laurence and Lewis Tierney of Bluefield, West Virginia, bought Fordson and renamed the operation Eastern Coal Company. The properties were sold to Pittson in 1968 and to Massey Energy in 1993. There is still some mining taking place at the No. 8 mine by an independent company.[131]

Today, Stone is a well-preserved company coal town and has been nominated as a Historic District. There are many houses and buildings from the 1910s and 1920s still in use. In the north end of Stone along KY 199 (which parallels the railroad tracks and Pond Creek) are the homes built for Pond Creek Coal's managers and office personnel. Many are nicely landscaped and have been well maintained over the years. The clubhouse built in 1912 was originally a two-and-one-half-story wood frame building with a series of dormers and a long porch that followed the L shape of the structure. The large right side of the clubhouse was destroyed years ago, and the building is now a white two-story home with a pyramidal roof and front porch.

The road winds through this residential area and continues into the commercial center. Two buildings, a brick neoclassical revival bank building with white columns and a wooden commercial structure, lie along Pond Creek.[132] Three large buildings, all three stories high, dominate downtown Stone and have been designated as landmarks by the Kentucky Heritage Council. The first floors have native-stone facades; the two upper stories are red brick. The first building has a large painted sign announcing "Red Robin Eastern Coal . . . Here American Workmen Mine, Prepare and Ship the World's Finest High Volatile Bituminous Coal." Originally the YMCA and theater, the building served as the town's post office until 1994. The main offices of the Pond Creek Coal Company were housed in the middle building. The main entrance is at street level, and five large window bays, capped by round arches and keystones, dominate the front of the structure. Eastern Coal continued to use the building for offices until the 1990s. The third building was the company store. A series of steps lead up to the main entrance and the display windows where customers were enticed with the latest goods shipped in on the railroad. On the road behind the three brick buildings is a large white wooden structure with a large porch. It was also a store.

Buildings constructed by the Fordson Coal Company after it acquired Stone, Kentucky, in the 1920s.

Right: This building served as the post office in Stone until the mid-1990s.

Below: Public Library at Stone, Kentucky, 1940. Norfolk & Western Historical Photograph Collection, Virginia Polytechnic Institute and State Universities Libraries.

The miners' homes lie just south of the company store. These basic two-story frame houses run alongside the tracks, set between the road and the creek. They are tightly packed together with little space for yards or sidewalks, and the front porches are just a few feet from the road. Coal companies have owned the houses until recently, so few have been significantly altered. Farther along KY 199 is Mullins Fork Road (on the right), which has large, brick suburban-style homes that were built for the company managers.

The industrial district of Stone is at the south end of town. Four buildings once comprised this district. West of the road, a three-story concrete powerhouse, with a railroad spur running through the middle of the building, generated electricity for the mining operations and town. A three-story concrete tipple was located just to the east of KY 199. A machine shop and warehouse were once located here. The remains of the tipple are located directly across from the warehouse building. The deep mine openings were on the hillside above the tipple near Love Branch Road.

There is a one-lane bridge and a cluster of miners' homes one mile farther as KY 199 runs through the communities of McAndrews, Pinsonfork, and McVeigh, where additional coal mining operations were developed along Pond Creek. McAndrews was established in 1912 and served the nearby Pinson Junction of the Norfolk & Western Railway. McVeigh was named for Pond Creek Coal Company's vice president Robert S. McVeigh.[133] There has been a renewed interest in Stone in the late twentieth century. The Pike County government and a local grassroots organization, Stone Heritage, have recently begun a project to restore the building complex in Stone and create a center for heritage tourism through the development of crafts, special events, and displays that celebrate the rich coal mining past of this community.

Following U.S. 23/119 West

At Pikeville, U.S. 23 and U.S. 119 meet and continue west as one four-lane highway into Letcher County, but the path of U.S. 23/119 has changed significantly in recent years. The original road paralleled the CSX rails past Escro and stayed east of the tracks until Dorton. From Dorton, it followed the rails south to Shelby Gap and west to Jenkins. The new four-lane U.S. 23/119 bypasses much of this area, entering Letcher County at Beefhide. To follow the CSX tracks more closely and see additional coal towns in this area, KY 1469 can be taken south from U.S. 23/119, through the Virgie and Jonancy area. At Dorton, old U.S. 23/119 continues to Shelby Gap and on to Jenkins. By either route, travelers can enter Jenkins, a major coal town of the early twentieth century, and continue on to many of the county's historic communities.

Letcher County (pop. 25,277) was formed from sections of Perry and Harlan Counties in 1842 and named for Gov. Robert P. Letcher. Three of Kentucky's great rivers have headwaters in Letcher County. The Levisa Fork of the Big Sandy, the North Fork of the Kentucky, and forks of the Cumberland all originate here. Letcher County was primarily agricultural in the nineteenth century until its rich coal deposits attracted the interests of investors in the 1890s. It was also the site of the famous Wright-Jones feud of the 1880s and the setting of a number of John Fox Jr.'s novels (see Big Stone Gap, Virginia). By the 1910s Letcher County had major coal company towns at Blackey, Jenkins, and McRoberts. The county boomed during the early twentieth century, but the 1950s brought decline and economic depression to many of its coal towns. Harry Caudill, a Whitesburg lawyer, brought Letcher County to the nation's attention with the publication of *Night Comes to the Cumberlands* in 1963. Caudill described the poverty and isolation of much of the region and helped inspire President Lyndon Johnson's War on Poverty. Like other Appalachian counties, Letcher County has experienced steady population decline in recent decades.[134] A visit to Letcher County can easily be combined with a trip into Wise County, Virginia, south on U.S. 23. The Appalachia and Big Stone Gap areas of Wise County have many historic coal towns and sites related to the mining industry.

Jenkins

Jenkins (pop. 2,401) is located at the base of Pine Mountain on U.S.23/119, approximately twelve miles from Whitesburg, the county seat. At an altitude of approximately fifteen hundred feet, it lies along the banks of Elkhorn Creek, which drains part of the Cumberland Mountains into the Russell Fork of the Big Sandy River. It was named for Baltimore financier George C. Jenkins, a major investor in the Consolidation Coal Company. Jenkins is Letcher County's largest former coal town and is still a well-populated, active community today. Developed by the Consolidation

Coal Company as a model company town in the 1910s, it once boasted a population of almost ten thousand people. Located at the site of the great Elkhorn coal seam, it prompted a mining boom in many nearby communities. Jenkins was a classic company town, built to the highest standards and promoted as a model mining community.

The site of Jenkins was developed later than some of the surrounding areas. Although Wise County, Virginia, was booming in the 1890s, the rough terrain around Pine Mountain and Elkhorn Creek made it almost impossible to reach the coal seams. Although there was some surveying for mineral wealth in the Jenkins area as early as the 1860s, it was not until the arrival of Richard M. Broas in 1881 that serious exploration began. Broas, a Civil War veteran and engineer from New York City, had extensive experience in mineral and coal ventures in western Pennsylvania and began conducting a survey of eastern Kentucky in 1881. He located two major coal deposits that were later developed by Consolidation Coal, the Miller's Creek seam in Johnson County and Letcher County's great Elkhorn seams.[135]

In 1883 Broas began a more detailed survey of the great Elkhorn seam, which follows Elkhorn Creek from its headwaters near Pound Gap, Virginia, eastward to present-day Elkhorn City, Kentucky. With the help of Nathaniel Stone Simpkins and his nephew John Simpkins of Massachusetts, he analyzed the purity of the coal, noting fine changes in its characteristics, and identified it as the highest quality coking coal. Broas located prime sections of the seam and purchased a large tract of land near the headwaters of Elkhorn Creek.[136] While they were surveying the Elkhorn area, coal mining was being developed in Wise County, Virginia. The Norfolk & Western Railway reached Big Stone Gap in 1890, and the Stonega mining district was opened a few years later. The Louisville & Nashville Railroad further connected Wise County with Middlesboro, Kentucky. But just north through Pound Gap, the Elkhorn district remained isolated. Broas was unable to persuade area railroads to penetrate this remote region. There were no real roads into the area, except a trail through Pound Gap, and the nearest railroad was almost sixty miles away. In frustration, Broas finally sold the lands to speculator John C. C. Mayo of the Northern Coal and Coke Company. Mayo traveled the area from Pikeville to present-day Jenkins (a route that later became known as the Mayo Trail), taking samples and purchasing additional lands along the way. His samples impressed mining engineers, and in 1910 Mayo sold 100,000 acres along the Elkhorn seam to the Consolidation Coal Company of Baltimore, Maryland.[137]

Prior to 1910 the banks of Elkhorn Creek were lightly populated with small farmers. While there were some opportunities in the timber trade, most residents were poor and economically isolated. Speculators such as Broas and Mayo took advantage of the limited economy by purchasing lands well below their fair market value. John Wright' s family owned much of the land along Elkhorn Creek in Letcher County. Wright sold his land to Mayo and then served as a middleman for Mayo. He later worked for Consolidation Coal, allowing them to acquire lands from family members and neighbors scattered along the creek and on the sides of Pine Mountain. Wright remained in Jenkins and became a feared lawman who claimed to have killed twenty-eight men during his career.[138]

Once lands were purchased, Consolidation Coal wasted no time in preparing the area for development. Even though the nearest railroad line was still thirty miles away, Consolidation Coal placed engineers and laborers at the site to begin construction of the town. Over the next year, Consolidation Coal poured its vast resources into creating a town in the rugged mountain terrain.[139]

Men traveled on horseback over the mountains from Virginia and began laying out the town. Equipment was hauled from the mouth of Elkhorn Creek to the site of Jenkins, a distance of more than thirty miles. A narrow gauge railroad used in the timber trade allowed for the transport of some materials, but it reached less than halfway to the site. Consolidation Coal leased the line, extended it five miles, and built a wagon road across the mountains another five miles. They also built a temporary independent line along Elkhorn Creek to transport supplies. The engines and boilers had to be hauled by wagons pulled by teams of up to twenty oxen to reach the site. Meanwhile, the company built a standard gauge line, the Sandy Valley & Elkhorn Railroad from Pikeville west into Letcher County. In October 1912 the first locomotive arrived in Jenkins.[140]

Engineers laid out the mine sites and a town for the hundreds of workers that would be needed. At first, timber cleared along the tracks was used to construct houses. Once the town was accessible by rail, Consolidation Coal contracted the Nicola Building Company of Pittsburgh, a company that designed and constructed houses for mining towns, to build more than one thousand homes in Jenkins and several adjacent communities. By

1913 more than two hundred houses had been finished and were regarded as the "best built anywhere." They featured fireplaces, closets, pantries, porches, and electricity throughout. By the early 1920s, most of the homes in Jenkins featured running water and sinks.[141]

Consolidation Coal mines were built with state-of-the-art equipment, electric lighting, and massive ventilation systems. Safety was a high priority.[142] The Jenkins mines were frequently featured in *Coal Age*, a promotional journal of the coal industry. There were three drift mines operating in the Jenkins area, and "True Elkhorn" was the purest, most uniform quality coking coal mined there. It was found in seams ranging from thirty inches to ten feet. Consolidation Coal trade-named this coal Cavalier Coal.[143] Despite a good safety record in the mines, Jenkins did suffer one major natural calamity. In February 1923 heavy rains caused a major mudslide at the No. 201 mine site near the tipple. The tipple was heavily damaged and the tracks and locomotives were buried in mud. Although there were a few minor injuries, no one was killed.

Consolidation Coal took great pride in its programs for miners and their families. It created an Employment Relations Department to oversee the "health, education, amusement, and recreation" of its workers. The company staffed its towns with professional nurses whose duties extended far beyond the hospital. They carried out instructional programs for families on hygiene, cooking, sanitation, first aid, and immunization. The medical staff was also responsible for the periodic sampling of local dairy and meat supplies to monitor their quality. In addition, they promoted group medical insurance policies available to Consolidation Coal employees that covered disability, medical, and death benefits. Nurses were assigned districts of the community and were encouraged to visit families and learn their individual needs.[144]

The Employment Relations Department supervised the town's educational and religious facilities as well. When Jenkins opened in 1912, a new school offered a high-quality education to miners' children. In addition to regular instruction, the school operated a kindergarten year round and held summer classes for students on domestic topics such as gardening, sewing, tools, and crafts. By the 1930s the Jenkins School District had more than three thousand students and hosted sports teams, a band, and literacy programs.[145] Religious activities were also encouraged, and within a few years Jenkins had Episcopalian, Methodist, Baptist, and Catholic churches.

Consolidation Coal employees enjoyed recreation centers and spaciously landscaped parks, lakes, and sports facilities. Elkhorn Creek was dammed and a fifteen-acre lake was created. A power plant was built at the north end of the lake, and parks and recreation facilities were built around the shoreline. The editor of the *Daily Independent* in Ashland, Kentucky, noted during a visit to Jenkins that the "clubhouse and lakefront look more like a summer resort than a mining camp."[146] The company promoted a wide variety of community activities through its agencies and religious institutions. Baseball and football teams competed with teams from Burdine, McRoberts, and Dunham.[147] Residents of Jenkins had all types of conveniences available, and the town had a YMCA, bakery, butcher shop, recreation center, schools, and a large company store. Consolidation Coal fought unionization but created a company association that arbitrated disputes and allowed employees to express their opinions about company policies. In order to discourage union activities, Consolidation Coal paid its miners wages above the union scale and provided a variety of benefits that were far superior to those available in many union mining towns.[148]

In the 1910s the outbreak of war in Europe dramatically increased the need for coal, and Consolidation Coal mines worked to break all production records. By then Jenkins had a large European immigrant, southern white, and African American population. The town became an ethnically diverse community with Italian, Yugoslavian, and Slavic neighborhoods. By 1920 Jenkins boasted a reputation as a modern city with a population of more than forty-seven hundred people.[149] Because of the large immigrant community, Consolidation Coal provided English classes to its workers and their families. Special seminars were also conducted to help immigrants assimilate to the American lifestyle. The YMCA was the center of these activities and was equipped to show lantern slide programs and motion pictures. The company had a professional staff who worked in the area, later serving the neighboring communities of Haymond, Fleming, and McRoberts. The variety of names on tombstones in the cemeteries at Jenkins, Burdine, and Dunham show the great number of ethnic groups that inhabited the town in its early years.[150]

Jenkins soon became east Kentucky's premier company coal town. Its mines, houses, and facilities were known throughout the Appalachian region. It had beautifully landscaped streets and a steadily growing population that topped 8,400 by 1930. It had its own dairy, butcher shop, produce department, and gardens. The Recreation Building boasted a theater, soda fountain, Western Union Office, and guest lodging.[151] The nearby coal camps of Dunham and Burdine were incorporated into Jenkins. Jenkins was one of 880 coal towns evaluated by the U.S. Coal Commission in the early 1920s, and it received a very good rating. The commission surveyed conditions in coal mining towns throughout the country and produced an eight-volume report of their findings in 1925. Jenkins was listed as 56 out of 713 company towns.[152]

Jenkins produced a number of famous people during its prime years. James Jones's best-selling World War II novel *From Here to Eternity*'s main character, Robert Lee Stuart, was based on Jenkins native Robert E. Lee Pruet, who was played by Montgomery Clift in the 1950s feature film. Country music singer Tennessee Ernie Ford's grandfather worked in the Jenkins coal mines for ten years and inspired his recording of the song *Sixteen Tons*.[153] Kenny Baker, a former miner, became one of the best known bluegrass fiddlers in the world and played in Bill Monroe's band. Like many towns in east Kentucky, Jenkins reached its peak at midcentury and began to decline during the 1950s. World War II temporarily boosted mining in the area, but by the late 1940s coal prices dropped. In 1946 Consolidation Coal sold Jenkins and surrounding properties, and it ceased to be a company town. Consolidation Coal continued mining around Jenkins until 1956 when it sold its properties to Bethlehem Steel, which continued working mines in Letcher County under its subsidiary, the Beth-Elkhorn Corporation. Bethlehem sold its properties in Jenkins in the late 1980s. As employment opportunities dwindled, people began to leave Jenkins. During the 1990s the town maintained a population of about twenty-seven hundred. Over the past thirty years, many of the major structures built by Consolidation Coal have been destroyed. The company stores, recreation centers, farms and dairies, offices, and tipples are all gone. In recent years the Jenkins Heritage Foundation has made an effort to preserve some of the surviving buildings. Even though many are gone, the remaining houses and public and commercial structures give visitors a good idea of what life was like in Jenkins fifty years ago. Hundreds of homes are still in use, and the town still has a flourishing commercial center.

East Jenkins was once Burdine, a separate town along old U.S. 23. A series of two-story clapboard houses appear along the creek through the town. They were built in the 1910s and until quite recently were almost all painted white, maintaining the original uniform appearance of this company town. Along U.S. 23 there is a small commercial center with two native-stone buildings from the 1920s that today house a music shop and an army surplus store. Uniform company houses continue along both sides of the road and are scattered on the hillsides on the approach to town. Many are two-story duplexes, each side having its own porches and entrance. Old U.S. 23 continues into town and becomes Jenkins's main street. The commercial center of Jenkins remains the heart of the town and has several historic buildings from the 1910s. Two large buildings lie along the south side of the road: the Jenkins School and the Jenkins Methodist Church. The two-story school was one of the original brick and masonry company buildings put up in 1912. It remained in use until the early 1990s when it was converted to office space for the Jenkins School District. The interior has undergone several restorations but still has much of the original woodwork in place. Next to the school is the Jenkins Methodist Church, another original structure dating from 1912. It is standing on the former site of the Wright family farm. This fine red brick church features stained-glass windows set in arched bay windows. Opposite the school are two brick buildings that once serviced equipment and mining machinery. Consolidation Coal's main mine, No. 204, was located beyond the railroad tracks.

Across from the machine shop is David A. Zegeer Coal and Railroad Museum, which is housed in the old railroad depot. The museum is named in honor of one of Jenkins's greatest boosters, David A. Zegeer. Zegeer came to Jenkins to work as an engineer for Consolidation Coal in the late 1940s. Like many other residents, he intended to work in Jenkins for a few years and move on. Instead, he ended up staying more than thirty years. After Consolidation sold its properties, he worked as division superintendent for Beth-Elkhorn. Zegeer later served as the assistant secretary of labor, Mine Safety and Health Administration, under President Ronald Reagan. Since retiring, he has spent much of his time working to preserve the memory and heritage of Jenkins. Zegeer, along with Marshall Prunty Jr., worked in the production of *The Birth of a Coal Mining Town: Jenkins, Kentucky*, a 1988 video documentary that tells the story of the town. It presents a very attractive picture of Jenkins and captures, according to Zegeer, the "pleasant history of a good town and good people." The museum is the latest project to promote the mining heritage of Jenkins. The former railroad depot stood empty for many years and was scheduled for demolition. Through the efforts of the Jenkins Heritage Foundation, the building was saved, funds were raised to restore it, and artifacts were gathered to be placed in permanent exhibits. The museum opened in 1998 and features displays of mining and railroad equipment and presents an overview of the development of the community in the twentieth century. Across the railroad tracks on the road leading up the hill is a small white wooden church with a simple steeple. Alongside are the rectory and a small garden. This is Saint George's Catholic Church, built to serve Jenkins's immigrant community in the early century.

Another good place to learn about the former grandeur of Jenkins is the Mary Jo Wolfe Library. Peggy Bentley, who has been very active in preserving the history of the town, eagerly shows visitors newspaper articles about the founding of the town and files of material on mining history. The library assisted with the video documentary and sells prints of historic photographs to raise money for library collections and local preservation efforts.[154] The most prominent company buildings that once stood in downtown Jenkins have been destroyed over the years. The company store, recreation building, and company offices are all gone. Even so, the center of Jenkins is still a lively place where a handful of stores and businesses continue to draw local residents.

A massive brick and stone powerhouse (demolished in 1988) that provided electricity for several of Consolidation Coal's towns once stood alongside the lake. The lakeside residential and recreational area can be reached by following the road that passes behind the Jenkins Hospital. This area was a popular spot for swimming and fishing and continues to be used for public and sporting events. On the wooded hillsides above the lake are comfortably shaded streets lined with homes that were built for Consolidation's managers, office personnel, and engineers. These two-story houses with spacious porches and yards were once considered the best Consolidation Coal had to offer. The company clubhouse was also located here, but it was lost in the 1970s. Although many of the original buildings in Jenkins are gone, the quiet neighborhoods and the beautiful lakeside area recall the pleasant, comfortable atmosphere that Consolidation Coal created for its employees in the 1910s.

Jenkins's rich heritage has also been preserved through the photographs of William "Pictureman" Mullins, a resident of Letcher County who photographed local townspeople for more than fifty years. Born in Virginia in 1886, Mullins moved to east Kentucky and established himself as a photographer. Working without electricity until the 1930s, Mullins developed his pictures in a darkened cabin with kerosene lamps and primitive homemade equipment. Most of his surviving works were made during the 1940s and 1950s when Jenkins was a thriving community. His pictures were simple; he seldom posed his subjects or arranged the settings. He specialized in family photos and events such as weddings, funerals, or baptisms. Mullins never seemed to be concerned about money; he charged modest fees and sold copies of his pictures for as little as fifty cents. Mullins died in 1969 at the age of eighty-two. In the 1980s Appalshop, a mountain arts and education center in Whitesburg, received more than four thousand negatives from his family and have made the collection an integral part of their efforts to preserve Appalachian culture. They depict the life of everyday people living in the mining and farming communities of the county and are an invaluable resource. The Mullins photographs have been exhibited throughout Kentucky and featured in many publications.[155]

A number of small communities were developed in the Jenkins area as an extension of Consolidation Coal's operations and other coal companies developed surrounding lands. They can be visited by following KY 805

The Consolidation Coal Company Store at Jenkins, circa 1920. Jenkins Collection, Mary Jo Wolfe Memorial Library.

west out of Jenkins. U.S. 23 and U.S. 119 divide at Jenkins, giving travelers two options for continuing a tour out of the Big Sandy River region. U.S. 23 heads south out of the valley into Virginia, allowing a visit to a cluster of coal towns near Big Stone Gap. U.S. 119 leads west out of the region into Harlan County, the site of two major coal towns that have undergone extensive renovation and preservation. A visit to the coal mining museum at Benham nicely complements a visit to any of the coal towns discussed in this guidebook.

KY 805 to Dunham

Dunham (pop. 100) lies along KY 805 two miles north of the junction and was developed in 1912 as an extension of the mining operations at Jenkins. Built by Consolidation Coal, it was named for A. S. Dunham, the company's auditor. Houses lined the ridge west of the railroad tracks, and the mine and tipple were located about one mile north along the tracks. Consolidation Coal built a preparation plant here in the 1940s and

Jenkins had its own meat market and dairy in the 1920s. Jenkins Collection, Mary Jo Wolfe Memorial Library.

Consolidation Coal Company marketed its Jenkins coal as Cavalier Coal. Jenkins Collection, Mary Jo Wolfe Memorial Library.

marketed its Jenkins area coal as Cavalier Coal. A large Cavalier Coal logo was placed on the side of the tipple.[156] Dunham had two stores. A main company store stood at the south entrance to town, alongside the tracks, and a small community store was located in the center of town. A small church and recreation center featuring a theater, pool, and restaurant once stood in Dunham. A small hotel provided board for single miners and newly arrived families. Today, a handful of houses from the era are visible just north of the highway.

Haymond

About three miles west of Dunham on KY 805 lies Haymond (pop. 400). A spur of the Louisville & Nashville Railroad opened the Haymond area to development in 1913, and the Mineral Fuel Company built a town and two mines at this site. The town bore the name of Thomas S. Haymond, Mineral Fuel's manager of local operations. By the end of 1914, almost 250 buildings were built along Potterfork, the site of Haymond. Plans were under way to expand the town when Mineral Fuel was absorbed into the Elkhorn Coal Corporation. The Elkhorn Coal Corporation also developed the nearby coal town of Hemphill and the commercial center of Fleming during 1913–15, located farther north on KY 805 and KY 317. By 1915 the Elkhorn Coal Corporation had absorbed Mineral Fuel and the Elkhorn Mining Company. Elkhorn worked the Haymond mines until the early 1950s when it ceased operations and leased its lands to the other coal operators. Haymond is more typical than Jenkins is of the many small company towns built in the early twentieth century. The mines sites and tipples were abandoned and demolished long ago, but there are a few commercial buildings scattered along KY 805. In the center of Haymond, a road leads east up the hollow, where tightly packed row houses appear much as they did when constructed in the 1910s. Although Haymond is no longer an important town, the Elkhorn Coal Corporation continues to be a major mining corporation in east Kentucky and has a headquarters in Prestonsburg.

Fleming-Neon

KY 805 can be followed west along the CSX tracks. About one mile north of Haymond, KY 317 meets KY 343 at Neon Junction. KY 343 passes through the center of Fleming-Neon and can be followed east to McRoberts. Fleming-Neon (pop. 840), a commercial center that once served surrounding coal camps, was originally two towns, the commercial center of Neon and the coal camp of Fleming. Today it is a quiet community, and the downtown commercial district remains much as it was in the mid-twentieth century.

Neon grew as a trading and commercial center in the 1910s, serving the coal towns of McRoberts, Fleming, Hemphill, and Haymond. It was located near the point where Wright's Fork and Potter Fork meet and was originally known as Chip, a name that may have been related to the timber trade in the area. Founded in 1913 and incorporated in 1916, the town changed its name to Neon, a name with several possible origins. Most likely, an early merchant brought a neon sign into the town. During the 1920s Neon was a booming town with almost one hundred businesses in operation. During this time, the Neon post office was established, along with four major restaurants, seven liquor stores, hotels, two cafes, two general department stores, apartments, a courthouse and jail, and specialty shops. A local newspaper, the *Neon News,* was published for twenty years until 1952. Two disasters hit Neon in the 1940s, destroying much of the original town. A major flood put the community under more than four feet of water in 1941, and a fire devastated the downtown commercial district in 1947. The two-story brick buildings that now line Main Street were built after the fire.

Neon's most famous business was a general department store that grew from the work of an itinerant Syrian peddler. Srur Dawahare sold general merchandise to residents of the surrounding coal towns, working the area around Jenkins after Consolidation Coal opened its mines there in the 1910s. He built a small wood frame building on Main Street in Neon in 1916. Private commercial ventures such as Dawahare's store did their best business in the later afternoons, after the company stores in the area closed. With three booming coal camps nearby, the store had plenty of customers and expanded in the 1920s. The Great Depression hurt, but

The once-thriving downtown commercial district of Neon.

by 1940 Dawahare was ready for expansion, opening clothing and merchandise stores in east Kentucky, and branch stores in Lexington, Louisville, Cincinnati, and Florence by the 1950s. Srur Dawahare's sons continued the business after his death in 1951, and Dawahare's remains one of east Kentucky's thriving retail outlets. The original building was replaced after the 1947 fire, and the company remained in business on Neon's Main Street until 1987, when it finally closed its doors after seventy-one years at the same location. Most of Neon's businesses have suffered the same fate as the population in the Fleming-Neon area and surrounding coal towns has continued to decline. Today there are only a few shops operating along the main street.[157]

Beyond Neon is Fleming, built in 1913–14 by the Elkhorn Coal Corporation on Wright's Fork of Boone Creek, a tributary of the Kentucky River. The town was named for George W. Fleming, Elkhorn Coal's first president. Elkhorn built houses for its employees, and by the time the mines were in operation in 1913, hundreds of people had moved into the community. The first train arrived in August and left with the first shipment of Elkhorn coal. A few years later the town had a hotel, bank, railroad depot, company store, butcher shop, machine shop, and Elkhorn's

Miners' houses along KY 343 in Fleming.

company offices.[158] Soon there were two main mines at Fleming, and the town continued to flourish. By 1920 there were churches, a fire department, jail, and an elementary and high school. The hotel, which served as a boardinghouse for employees, provided many commercial and recreational services for the community. The lower level featured a movie theater, barbershop, pool hall, post office, and doctor's office. The First National Bank opened its doors in 1921 and served the financial needs of miners and their families until 1937. A new school, Fleming High School, was completed in 1928.

Elkhorn mine No. 302 operated at Fleming until the 1950s. The town remained a headquarters for Elkhorn Coal Corporation until the 1950s, when it ceased mining operations in Haymond and Hemphill. Today Fleming lies between Neon and McRoberts and is difficult to distinguish as a separate community. There are several small commercial buildings along the road and several abandoned mine sites hidden in the nearby hills. The houses that extend from Fleming to McRoberts are two-story wooden frame front-gabled houses with front porches and brick chimneys strung out along KY 343. They are tightly packed together, with just a narrow sidewalk separating the front porches from the road.

McRoberts

McRoberts (pop. 921) lies about three miles east of Fleming-Neon along KY 343. It was another coal town built as an extension of Consolidation Coal's operations at Jenkins. Located along Wright's Fork of Boone Creek, a tributary of the North Fork of the Kentucky River, the town was constructed in 1912 along a spur of the Louisville & Nashville Railroad. The L & N Railroad brought the supplies and materials needed to construct the town and delivered mail from Lexington via Hazard. Passenger service was added later. The new town was named for Samuel McRoberts, a New York City financier who later became the company's director in 1918.

The construction of the town began shortly after Jenkins, and by early 1912 two sawmills were cutting lumber to build houses. The Nicola Building Company of Pittsburgh later built many of the houses in McRoberts. The town was laid out along a wide boulevard divided with a landscaped green. Most of the houses were two-story wooden frame buildings. The majority were built for two families, with each family having three rooms downstairs and two upstairs. Each house had a front and back porch, as well as a coal house and an outhouse, and a water pump was located in front of every third house. The position of a worker in the company determined what type of house he and his family lived in. Houses built for foremen, managers, and supervisors often had basements, bathrooms, electricity, and telephones. Both Jenkins and McRoberts drew far more people than they could at first accommodate. A hotel and boardinghouse served new arrivals, single men, and visiting company officials.[159]

McRoberts offered its residents many of the conveniences and services that were available at Jenkins. A large company store was placed in the center of town with a meat market next door. Several smaller stores appeared later. By the 1920s there were separate shoe stores, barbershops, garages, filling stations, and several restaurants. A church was built by the company to serve the mining community. By the 1930s two schools operated at McRoberts, an upper school and a lower school, serving different grade levels. In 1918 the company assigned a doctor and nurse to the town's new clinic. Recreation was also provided for the residents of McRoberts, including a recreation center with a large ball park, tennis courts, a theater, a soda fountain, and a tobacco and candy stand. A pool room, located downstairs in the recreation building, sold beer on Saturday nights.

Two large mines (Consolidation No. 210 and No. 211) were located at McRoberts, and a central tipple was constructed by 1913. By the 1940s there were six mines in the McRoberts area and four tipples. McRoberts was never as large as Jenkins, but hundreds of men worked in the mines. Rows of company houses spread out from the town along the tracks to the tipple. When mine No. 213 was built, the area became known as "13 Row." Many of the African Americans lived in the Tom Biggs area of McRoberts, where there was a separate school, stores, and recreation center. The Tom Biggs area was also the site of many service facilities, including a motor barn, a blacksmith's shop, a water plant, and a horse barn for company horses used in the mines. A lodge hall housed the post office, union meeting area, doctor's office, and scrip office.

Jenkins and McRoberts were both major Consolidation Coal Company towns but were separated by a mountain and miles of winding mountain roads; however, they were connected underground. One could enter mine No. 214 at McRoberts and walk to Jenkins, exiting at the No. 207 mine. This route was sometimes used for emergencies to transport people to the main hospital in Jenkins.[160] McRoberts prospered until the

1940s when Consolidation Coal sold its properties in Letcher County. Since the 1950s several coal companies have worked the area around McRoberts but have employed far fewer people. McRoberts is much smaller today and most of the original company buildings have been lost over the years. Many were demolished during the 1950s, including the clubhouse and recreation building. In fact, it is difficult to distinguish McRoberts as a separate town, the row of houses extending from Neon to McRoberts and the lack of remaining company buildings and mine sites can blur the three communities. So much of McRoberts has been demolished or overgrown, making it difficult to imagine what a large and vibrant community it once was.

There are only a few sights to see today in McRoberts. The Church of Christ sits on the former site of the recreation building. The lodge hall was destroyed to make a parking lot for the Old Regular Baptist Church. The

Veterans' monument in the center of McRoberts, Kentucky.

company store was taken over by a private commercial chain but was later lost in a fire. The center of McRoberts was located near the Missionary Baptist Church, which was originally a community church when organized in 1925. Near the church is a monument to the many veterans who served in the armed forces, erected in 1942 by the UMWA. The white marble column stands in the center of the shaded boulevard and honors local residents who served in the Second World War and all of the following conflicts, including the Persian Gulf War. The boulevard still runs the length of the town. The upper school building had its second floor removed and now serves as a private residence. The nurses' residence and the city hall are still standing. An unused boardinghouse is still standing, as is the two-story superintendent's house near the center of town.

McRoberts's population has dwindled in recent years, although many retired miners still live in the town. McRoberts was one of the few area communities with a substantial African American population, and some of their descendants still live in the homes along KY 343. A reunion of former residents and people with family ties to McRoberts takes place every two years in the ballpark, usually in August. An active community group organizes fund-raisers for the reunions and helps maintain the community cemetery.[161]

Seco

Seco (pop. 175), located on KY 805 just past Neon junction, can be reached from U.S. 119 about five miles northeast of Whitesburg. Seco lies on Boone's Fork, about one and one-half miles from where it meets the North Fork of the Kentucky River. The town was developed by A. D. Smith and Harry Leviers of the Southeast Coal Company, who named the town after the company, SECO. The town and post office were constructed in 1915. The Southeast Coal Company was a sister company of the Northeast Coal Company, which operated in the east Kentucky counties of Floyd and Johnson. Both were known for providing quality housing and facilities for their employees. Unlike Consolidation Coal or Elkhorn, these two companies only hired white mountaineers and a small number of immigrants, excluding African Americans from their workforces. The company built churches, a boardinghouse and clubhouse, and a large company store and hospital. Regular freight service kept the store well stocked. By the 1920s three mines were opened, and Seco drew hundreds of miners and their families. The town prospered during the 1920s, but the Great Depression brought problems. By the end of the decade, the mines had reopened and prosperity returned.

Local residents recall life in Seco as a generally pleasant experience. The company provided good services for miners and their families. Long-time resident Opal Jeanne Tuggle compared life in Seco to life on a farm. There were many isolated, wooded areas, pasture lands for livestock, gardens, and open spaces. During the Great Depression of the 1930s there was a community spirit that helped people survive. SECO was regarded as a fair company, and unions had little influence. There does not appear to have been any major labor disputes in the town. The mines were run well and built with safety in mind. Residents do not recall any major disasters in the mines at Seco.[162] The town prospered until the 1950s when the company suffered from the decline in coal markets. The mines at Seco closed in 1957, and although the company offices remained open for five more years, Seco was no longer a company coal town. By the 1960s properties were sold to individuals and Seco became an incorporated community. Although the mines were occasionally worked, most residents of Seco found employment in surrounding areas.

From U.S. 119, a small bridge crosses Boone's Fork and leads into the center of town. The United Methodist Church is located at the entrance to the town. This long, narrow white wooden structure was one of the last built by SECO in 1939 and is still in use more than sixty years later. There are about fifty houses originally built by SECO that line the roads through town. These are basic, front-gabled two-story shotgun houses with small porches and varied pitched roofs. They are narrow and fit fairly close together on the small lots throughout the town. Most are in good condition and have been well-maintained and modified over the years.

The most prominent building in town is the former Seco Company Store, located at the far end of the square along the railroad grade. It is a three-story frame structure with two side additions. The front porch and sloping roof cover the full length of the facade. The store building is currently undergoing major renovations by its owners, Chris and Sandra Looney, who plan to reopen the building as a winery, tourist site, and bed and breakfast. Even though there were several major fires in the 1920s, the store's basic structure has remained intact. Many of the original interior wooden beams, shelves, and cabinets remain, and there is a large root cellar where fruits and vegetables were kept. The store offered a great

The South East Coal Company store at Seco, Kentucky.

variety of services, products, and conveniences and served as the social center of the community. On display is a ledger from the 1950s that indicates miners' purchases, credits, and debts. It is worth noting how few of the miners appear to have a debt to the company store.

Beyond the store, along the old railroad bed, the road passes the club-house, a three-story wooden frame building that once served as a recreation center and boardinghouse. Unfortunately, it is structurally weak and may be beyond the point of restoration. The road between the club-house and the store leads to the two main mine sites. Before reaching the mines, the road winds through Boss Hill, an area where company managers and supervisors lived with their families. The houses are noticeably larger, better built, and have more spacious grounds. Across the street is a cemetery set aside for company officials and their families.

The miners' bathhouse, located where the road divides, is a wooden and corrugated metal structure with a concrete interior. Miners bathed and changed clothes here at the beginning and end of their shifts. The interior of the bathhouse contains much graffiti left by miners over the years. Beyond the bathhouse are two main mine sites, some abandoned equipment, the old tram route, and several slate dumps. Although all these sites are located on private property, arrangements can be made through the store to view the clubhouse, bathhouse, and mine sites.

Directly across from the store is the site of the first houses in Seco and the former location of the company hospital, which was demolished in the 1950s.

Currently, parts of Seco are being restored and developed for heritage tourism. A good time to visit is on Memorial Day weekend, when a Miners' Memorial Celebration draws people from all over who have ties to the community.

Whitesburg

Whitesburg (pop. 1,636) is located about fourteen miles west of Jenkins on U.S. 119. Named for Daugherty White, a state legislator who helped develop the area in the 1840s, Whitesburg is a regional commercial center and the seat of Letcher County. In the late nineteenth century, it grew as timber was floated down the Cumberland and Kentucky Rivers and was milled in Whitesburg. Before the railroad arrived in 1912, Whitesburg was the only town in Letcher County. During the 1920s it served the surrounding coal towns of the county with commercial and government services.[163]

While most of the citizens of Whitesburg go about their daily business, a small group of artists, activists, and writers have transformed Whitesburg into a center for the arts and humanities over the past thirty years. The Appalshop Community Arts and Media Center is located at 306 Madison Street, next to the North Fork of the Kentucky River. It features galleries, live performances, a film and video production center, and a radio station. The center concentrates on depicting the history, culture, and social problems of Appalachia and hosts workshops highlighting local performers. Appalshop is best known for documenting the vanishing regional cultures of Appalachia and playing an activist role in environmental issues. People from all over the world visit Whitesburg to experience Appalshop and attend its annual celebration of Appalachian culture, Seedtime on the Cumberland, which takes place the first week of June.

Whitesburg was also the home of Harry Caudill, whose shocking account of the decline in the coal industry and the impoverished conditions in many area towns brought national attention to the plight of

Appalachia and was partly responsible for President Lyndon Johnson's War on Poverty. In *Night Comes to the Cumberlands* (1963), Caudill described the collapse of coal in the 1950s, the struggle for survival among miners and their families, the development of the welfare system, and the ravages of mining on the landscape and environment. Letcher County's branch library on Main Street is now named the Harry M. Caudill Memorial Library.

To learn about local politics and culture and to hear some authentic mountain music, tune into Appalshop's community-operated radio station, WMMT at 88.7 FM, while touring the region. While in Letcher County, pick up a copy of the Letcher County *Mountain Eagle,* a democratic activist newspaper with "Speak Your Piece," a feature that has become a local institution. For a breathtaking view of Whitesburg and the surrounding area, follow U.S. 119 west as it climbs Pine Mountain. Near the summit is a view of Whitesburg, about 1,800 feet below.

Blackey

Blackey (pop. 153) is located fifteen miles north of Whitesburg on KY 7, on the banks of the North Fork of the Kentucky River, alongside the CSX tracks. Originally called Indian Bottom, the town was later named for a local resident, Joe Brown, whose nickname was Blackey. When passenger and freight service was established in 1913, the Louisville & Nashville Railroad called its depot Blackey, and the name of the town was formally changed in 1919.[164] In the 1910s, as the wartime demand for coal grew and coal companies opened up Letcher County to mining, Blackey began to grow and prosper. The area around Blackey was mined by four major companies in the 1920s: the Blackey Coal Company (1917–25), Marion Coal Company (1919–28), Rockhouse Coal Company (1919–29), and Elk Creek Coal Company (1919–24). Blackey became a commercial center serving eleven surrounding coal towns and almost fifteen hundred miners. It was soon the second largest city in Letcher County, with a mayor and city council. Local merchants built houses to rent to workers from the nearby mines. As a stop on the L & N Railroad, the town attracted miners and their families for shopping and recreation. Among the many businesses in the town were grocery stores, dry goods and clothing stores, and hardware and drug stores. Directories listed restaurants, lodges, shoe shops, hotels, and pool rooms. There was an automobile dealership, the

Blackey State Bank, and, the most popular attraction, the Blackey Theater. A nine-chair barbershop did a thriving business on weekends, and sporting events were held at the school, which had tennis courts. The town's population grew steadily to more than seven hundred residents by the early 1930s. Weekend visitors added several hundred to these numbers.[165]

Although prosperous, Blackey suffered from several disasters that destroyed much of the town in the 1920s. In December 1926, a devastating fire swept through town, and frozen water lines hindered efforts to control the flames. Although dynamite was finally used to blast areas around the fire, the next morning less than half of Blackey remained. The following May, as residents were rebuilding, the worst flood in Letcher County history wiped out roads and bridges and placed the town under several feet of water. A second fire in November destroyed many of the remaining structures.[166] The declining coal market of the late 1920s and the onset of the Great Depression almost completed the destruction of the town. The surrounding mining communities collapsed as most of the mines closed. By the mid-1930s, the community was in trouble. The area coal camps were abandoned and many had people fled the area. Without customers, most of the retail stores had closed and only a small grocery remained. The collapse of the nation's economy affected local finances as well. The Blackey State Bank did not survive the Great Depression, closing its doors in the 1930s. Despite the efforts of several New Deal programs around Blackey, the town's days were numbered. The town government dissolved and Blackey was almost a ghost town.[167] Although a few hundred people remained through the 1950s, Blackey never recovered. By 1960 it was a quiet, isolated settlement that served a few truck mines nearby. In 1979 there was a brief recovery when Hollywood producers used Blackey in some location shots for *Coal Miner's Daughter*, the story of country music singer Loretta Lynn. The production provided small parts for a few local residents and gave the town a brief burst of fame.[168]

From KY 7 a steel bridge leads across the river to the center of Blackey. The Blackey Library on Main Street has several files that will help acquaint the visitor with the history and former importance of the town. There are only a few structures remaining today. Up the street from the library is the remains of Blackey's booming downtown commercial district. Beyond the vacant lot and small mobile home is a two-story brick structure that was the Blackey grocery, with its fading painted lettering still visible on the east side. Beyond that is a two-story flat-roofed stone and brick building that was the Blackey State Bank in the 1920s. Across the railroad tracks is the Presbyterian Church, also from the 1920s, and two large two-story homes.

For many years, the main attraction in Blackey has been C. B. Caudill's Store on KY 7 at the entrance to town. Opened by C. B. Caudill in 1933 and expanded in 1940, the store served the Blackey area for more than six decades, well into the late 1990s. This balloon-framed, steel-supported wooden building has a back-sloping metal roof that acts like a false front, making the one-story structure appear taller. A long white oak porch stretches more than one hundred feet across the front of the store, facing KY 7. The store is typical of independent rural general stores found throughout the region in the 1920s and 1930s. It carried a wide variety of goods unavailable to local residents unless they traveled to Whitesburg, seventeen miles away. The building was divided into two areas, one served as the retail area, the other as living quarters. A centrally placed stove warmed both the store employees and patrons during the cold winter months. Perhaps the most important area of the general store was the front porch, where people could spend the leisurely hours of the afternoon visiting with neighbors and catching up on news and gossip.[169]

Joe Begley worked for Caudill in the late 1940s and married Caudill's daughter Gaynell. When they took over the store in 1966, they kept the original name intact. The Begleys operated the business until 1997, and it became a local institution, looking more like a museum than a store. The Begleys gathered mining artifacts, local history materials, and a great variety of obscure and rare products. There were examples of company scrip, boots, lights, and assorted tools. The store had no formal organization, and visitors had to hunt for some of the more interesting items, such as hair restoration creams and exotic mining tools. Because of the great wealth of artifacts and information, the building has recently reopened as a history center and museum.

Joe and Gaynell grew up in east Kentucky when people still traveled by horseback and river steamboats. Joe left the mountains for factory work in New England and then served in the U.S. Navy during the Second World War. Gaynell attended Berea College and did graduate work in Chicago. They met in Connecticut and married in 1942. Later they returned to the mountains and worked in a variety of jobs and places. In 1966, after the death of C. B. Caudill, they returned to Blackey to run the store. During their travels throughout Appalachia they became aware of the great damage caused by strip-mining and the many problems facing miners and their families in the boom-bust cycles of the coal industry. Inspired by the works of Harry Caudill, they became active in citizens' leagues and community action projects. They fought strip-mining projects and worked for environmental protection for the area's many mountain communities. Joe became one of east Kentucky's best-known activists during the 1970s. He was instrumental in lobbying for the 1977 Surface Mining Control and Reclamation Act (more commonly known as the 1977 Strip Mining Act), which requires coal companies to restore stripped lands to their original contour. President Carter invited Joe and Gaynell to the White House for the signing of the bill in 1977.

The store became a center of activism as well, and the Begleys became celebrities as they entertained politicians, students, activists, and journalists. They were featured in Studs Terkel's *American Dreams: Lost and Found* (1980) and later in his 1997 work *Coming of Age*. Terkel was impressed with Joe Begley's commitment to Blackey, describing him as a "ambling, gaunt, Lincolnesque figure" who lived in "what appeared to be a ghost town." He portrayed the Begleys as having a kind, patient manner and maintaining close personal relationships with their customers. He quoted

Joe Begley behind the counter of the C. B. Caudill Store, 1996.

Gaynell explaining that "a transaction here is not entirely economic. Its a matter of friendship and socializin' for a minute. That's as important as getting that quarter."[170]

The Begleys also campaigned for better health care and rural services. Most recently, Joe was involved in the fight to preserve environmental legislation under attack in the Congress in the late 1990s. Although Blackey is no longer an important mining town, the Begleys helped keep the name alive by preserving much of the region's past and being actively involved in mining issues.[171] For their efforts over the years, the Begleys recently received the Helen Lewis Community Leadership Award, given by the Mountain Association for Community Economic Development, based in Berea. They were cited for their commitment to public service issues and their work in preserving the mining heritage of Letcher County.

After the store closed in 1997, the Kentucky Arts Council and Appalshop granted funds to reopen the facility as a store and museum. It was placed on the National Register of Historic Places and named the C. B. Caudill Store and History Center (Joe thought the word *museum* would make people think it was boring). The center is open by appointment to visitors and has hosted teacher workshops and school fieldtrips. Student interns from area colleges have helped gather historical materials and prepared the National Register application. Joe and Gaynell helped with the project as long as they could, but on 27 March 2000 Joe Begley died at the age of eighty-one. Just a few hours earlier he had been on the telephone trying to help block one of the latest strip-mining projects planned in the county.[172]

Lilley Cornett Woods

Located southwest of Blackey on KY 1103, off KY 7, the Lilley Cornett Woods offers visitors a unique opportunity to view the Appalachian wilderness as it was before industrialization. The land lies along Line Fork Creek, a thirty-eight-mile tributary of the North Fork. It is one of the few virgin forests still in existence in the United States. It has never been subjected to the timber trade, mining, or land development and has remained virtually untouched. Botanists claim the area represents an unbroken succession of plant and animal life going back as far as the Pleistocene Epoch, seventy million years ago. Trees as old as four hundred years climb uninterrupted, shading the floor of the forest and protecting its delicate balance of vegetation. The Lilley Cornett Woods is 554 acres of old-growth forest and is classified as a mixed mesophytic forest region in an untamed state. The upper slopes of the hills are filled with oak, beech, maple, hickory, black walnut, and hemlock. The shaded forest floor below supports a unique blend of wild flowers, fungi, ferns, and assorted vegetation. Animal life varies with the seasons.

Lilley Cornett purchased the land shortly after World War I and vowed to protect it from any outside development or environmental abuse. He fought off thieves and squatters, refused great offers from timber and mining companies, fought forest fires, and allowed the natural environment of the forest to flourish while all around it land was being destroyed. The lone exception was during a devastating blight of chestnut trees in 1929. Cornett permitted a few trees to be cut, hoping to save many others. He used wood only from these disease-ravaged trees to make the few fences found on his farm. Cornett became a little-known folk hero among environmentalists, and after his death his family continued to protect the land and sought a sponsor to conserve the forest for future generations. In 1969 the forest was purchased by the state of Kentucky, which maintains it today. The park is open from spring to late fall. To preserve the forest, visitors must take a guided tour.[173]

Harlan County and the Big Stone Gap Coalfields

6

South on U.S. 23

For travelers leaving the Big Sandy River Valley and traveling south into Virginia on U.S. 23 there are several interesting sites in Wise County, Virginia (pop. 40,123), just across state line from Letcher County. A cluster of coal towns around Appalachia, Virginia, and Big Stone Gap, along with a series of museums and historic sites, are all places to continue exploring the coal mining heritage of southern Appalachia.

U.S. 23 crosses into Wise County, a major center of coal mining today. Coal trucks regularly travel this route, and modern mining sites can be seen along the way. Wise County, Virginia, was formed in 1856 from parts of Lee, Russell, and Scott Counties and was named for the state's governor Henry Wise. The county seat was incorporated as Gladeville in 1874, but the name was later changed to Wise. The county's economy was based on farming and logging until the 1880s when coal mining drew the attention of investors. Confederate Gen. John Daniel Imboden brought the great natural resources of the area to the attention of eastern investors and helped organize early timber, iron, coal, and railroading ventures. Pennsylvania anthracite coal companies developed the Big Stone Gap coalfields in the 1880s, and branches of the Norfolk & Western Railway soon penetrated the area. By 1890, with the help of Gen. Rufus Ayers and John Taggart, the Virginia Coal and Iron Company began developing coal mines around present-day Stonega. The first shipments of coal left Wise County in 1892, and by 1900 the Big Stone Gap coalfields were some of the most productive in southern Appalachia. When Stonega Coke and

Coal, a lessee of Virginia Coal and Iron, absorbed several Virginia companies, it became the primary builder of company coal towns in southwest Virginia, establishing ten by the 1920s. As the Norfolk & Western Railway approached the area in 1889, the city of Norton was laid out. Appalachia was born when two railroads, the Louisville & Nashville and the South Atlantic & Ohio, met there in 1891.

Coal has dominated the economy of Wise County throughout the twentieth century although employment opportunities have been greatly reduced in the past decades. Recently there have been several projects to diversify the economy of Wise County. Several industrial parks and light telecommunications projects have added jobs, and Norton and Wise have become retail shopping hubs for surrounding communities. The Norfolk-Southern Railroad continues to play an important role in the shipment of coal and operates a large classification yard at Norton.

Wise County offers museums that preserve the coal mining past and a number of coal mining towns clustered around Big Stone Gap and Appalachia. A visit to Wise County can easily be combined with trips into Letcher or Harlan Counties just north in Kentucky.

Big Stone Gap

A regional commercial center and the site of several important landmarks and museums relating to the coal boom of the early twentieth century, Big Stone Gap (pop. 4,856) was originally called Three Forks, because three forks of the Powell River meet here. The town was laid out by engineer John Nadar in the early 1880s. As timber and mining operations developed, the town became known as Mineral City, but in 1888 the name was changed to Big Stone Gap. When the South Atlantic & Ohio Railroad reached the town in 1891, an old iron furnace was dismantled in St. Louis and transported to the town. By 1892 companies were developing the coalfields around the town and company personnel and investors were moving into Big Stone Gap. The surrounding area, which was developed primarily by the Virginia Coal and Iron Company and its lessee Stonega Coke and Coal, is known as the Big Stone Gap coalfields.[1]

The Interstate Railroad was organized in 1896 by the Virginia Coal and Iron Company. Originally it was built to take coal from mines to branch lines of the South Atlantic & Ohio Railroad and the Louisville & Nashville. After 1904 it was expanded and absorbed several small lines to serve the many coal towns developing around Appalachia and Big Stone Gap. Until the 1960s the Interstate Railroad was the main hauler of coal in the region.

The Interstate 101, an early-twentieth-century passenger car, was built by the Pullman Company in 1870 and purchased by the Interstate Railroad in the 1920s. For more than thirty years it was used as the railroad president's private car and for special excursions. It originally contained an observation room, two state rooms, a kitchen, and dining facilities. The car was retired in 1959, but in 1988 it was restored to its present condition and returned to Big Stone Gap. Today it houses the Tourist Information Center, located south of town on Business U.S. 23.[2]

Three local sites celebrate the works of John Fox Jr. (1863–1919), one of America's most popular novelists at the turn of the century and a former resident of Big Stone Gap. Fox wrote many books and short stories but is most famous for *The Trail of the Lonesome Pine* (1908) and *The Little Shepard of Kingdom Come* (1903). His works often depicted the opening of the southern Appalachia area to mining in the late 1800s and the transition from simple mountain life to a modern industrial society. Fox was born

The Interstate Railroad president's private Pullman car, now a Tourist and Information Center in Big Stone Gap, Virginia.

near Paris, Kentucky, and studied at Transylvania College and Harvard University. In 1888, while building a reputation as a newspaperman in New York City, he came to Wise County with his two brothers, who were coal operators in Tennessee. Together they were involved in land speculation in the area of Big Stone Gap. During his work in the field, Fox became fascinated with mountain life and left the coal business to concentrate on his writing.[3] He often lived in Big Stone Gap but spent much of his time traveling and pursuing his career as a journalist. Fox wrote short stories, news reports, and served as a foreign correspondent in the Spanish-American and Russo-Japanese wars. He had a sophisticated circle of national political and academic friends, including President Theodore Roosevelt. For several years Fox was married to Metropolitan Opera star Fritzi Scheff. She was so infatuated by Fox and his description of life in the mountains that she divorced her husband, married Fox, packed her bags, and arrived in southwest Virginia with her servants. For several years they were local celebrities in Big Stone Gap. The marriage did not last long, as Scheff soon found life in Big Stone Gap to be much less exciting than the one she had left in New York. After their divorce in 1913, Fox traveled the world once again. He later returned to Big Stone Gap, where he died in 1919.

In *The Trail of the Lonesome Pine* (1908), simple Virginia mountain culture is contrasted with the modern outside world. Jack Hale, a mining engineer, falls in love with a mountain woman, June Tolliver. He persuades her to leave her rustic home to receive an education, and she moves to Big Stone Gap, Louisville, and finally New York City. The work is full of detail about local mountain life at the turn of the century, including bean stringings, corn shuckings, log cabins, and moonshining. *The Trail of the Lonesome Pine* became the first American novel to sell more than one million copies. It eventually sold over two million. A successful stage version of the work was performed at the New Amsterdam Theater in New York, and several movie versions were produced by Cecil B. DeMille in 1916, 1922, and 1936. The 1936 version featured Fred MacMurray, Henry Fonda, and Sylvia Sidney.[4]

On Shawnee Avenue is the John Fox Jr. Museum, located in his former home, which is listed on the National Register of Historic Places. The two-story house was originally a small cottage residence for the Fox brothers but was added to over the years. It is now a twenty-two room two-story frame dwelling. It sits on a stone foundation and is covered with dark brown wood shingles and several brick chimneys. This is where John Fox spent his most productive years writing his best-known works.[5] The museum contains original furnishings and memorabilia from the family. The Trail of the Lonesome Pine State Outdoor Drama is located on Clinton Avenue and presents a theatrical version of Fox's *Trail of the Lonesome Pine* during the summer months. The outdoor drama, featuring mountain music and local talent, was developed by Earl Hobson Smith and has been performed in Big Stone Gap since 1964. A third site is the June Tolliver house, also listed on the National Register, on Clinton Avenue. The nineteenth-century house is named for the *Trail of the Lonesome Pine* heroine, June Tolliver, whose character was based on a young girl named Elizabeth Morris. Morris lived in this house when she came down from the mountains to attend school. The home features a craft shop, a collection of Fox's works, and nineteenth-century furnishings.

Big Stone Gap also has two museums that depict the development of coal and iron ore mining in the area. The Southwest Virginia Museum on First Street is located on a knoll known as Poplar Hill and is the former residence of Gen. Rufus Ayers, who was instrumental in attracting capitalists and railroads to southwest Virginia. Ayers was the son of a prominent Virginia planter from Bristol. After the Civil War, he studied law and invested in lands that held rich coal and iron deposits. He then promoted the southwest Virginia area to northern speculators. Ayers organized the Appalachian Steel and Iron Company at Big Stone Gap and the Virginia, Tennessee and Carolina Steel and Iron Company, and he managed the Wise Terminal Railway. By the 1880s Ayers held more than ten thousand acres of coal lands in Buchanan and Tazewell County. Ayers managed Virginia Coal and Iron Company properties and helped establish the first coal mining operations at Stonega.[6]

Ayers was one of a handful of local elites who prospered from the coal boom, and the museum reflects the lifestyle of these successful investors. The two-and-one-half-story house was constructed in 1888 and took seven years to complete. It sits on a prominent elevated lot and is surrounded by a native stone wall that matches the limestone and sandstone exterior of the building. The house has a series of stone steps that lead up to the front entrance. There are spacious windows, dormers, and four large chimneys. Inside the museum are displays that depict the history of southwest Virginia, including its role in the Civil War, the development of the coalfields, and local geology. The second floor features artifacts of everyday life in Big Stone Gap in the early years of the twentieth century,

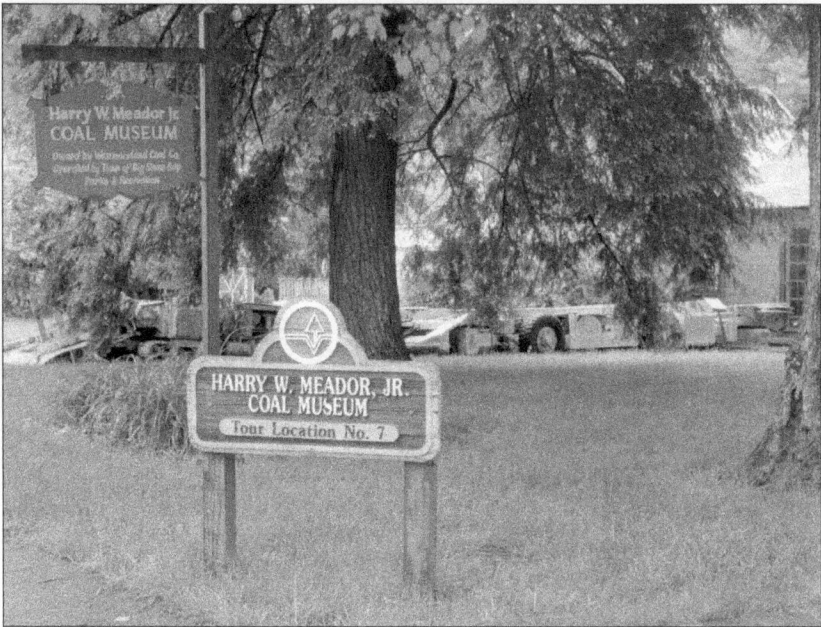

The Harry W. Meador Jr. Coal Museum is a good place to get acquainted with the history of the Big Stone Gap coalfield.

including mail order catalogs, phonographs, railroad memorabilia, and household items.

The Harry W. Meador Jr. Coal Museum is located on Shawnee Avenue in John Fox Jr.'s former library and study. Meador, for whom the museum was named after his death in 1981, started out as a miner and ended up as vice president of coal development for the Westmoreland Coal Company in the 1970s. He was instrumental in organizing the museum, which has extensive displays of coal mining equipment, tools, and memorabilia, and helped plan exhibits and tours. The collection includes a three-foot hand-carved wooden statue of a coal miner, a large collection of photographs showing many of the coal mining operations as they existed in the Big Stone Gap fields, and exhibits of medical equipment from the company hospital, tokens from coal cars, and pay vouchers. On the museum grounds are signs from the Stonega Coke and Coal Company and large mining equipment, including a Jeffrey Shuttle car that transported coal to a conveyor belt, a Lee Norse Continuous Miner, and a 1920 model of a Goodman Cutting machine used to undercut coal from the face of mines.

Beautiful Victorian-era homes line the streets of Poplar Hill in Big Stone Gap, Virginia.

On Wood Avenue is a small park dedicated to the miners of the Big Stone gap coalfields. Miners' Park stands on the former site of the elegant Monte Vista Hotel, once an important social center for local elites.

Big Stone Gap is an interesting town to explore because of its Victorian-era homes and residential neighborhoods. Owners of railroad and mining companies, such as the Wentz and McCorkle families, built these beautiful homes in the 1880s in the area known as Poplar Hill, and some of their descendants still live there today. Farther south, on Wyandotte Avenue, are more beautiful homes and churches from the 1890s.

Appalachia

Appalachia (pop. 1,839) is located on U.S. 23, just north of Big Stone Gap. In 1891, during the land and coal boom that began in neighboring Big Stone Gap, the Louisville & Nashville Railroad made a junction here and named it Appalachia. The South Atlantic & Ohio Railroad reached Appalachia a short time later. The Interstate Railroad laid tracks connecting the

Appalachia junction with its many coal towns. The town grew quickly in the late 1890s as a railroad junction and regional commercial center. Several general merchandise stores were opened by 1898 and many other soon followed. Major hardware, furniture, and department stores followed. Within a few years, there were restaurants, theaters, cafes, and saloons all along the main street.[7]

By 1906 the main streets were laid out and Appalachia was incorporated. It served as a commercial center for the more than thirty thousand people who lived in the nearby coal towns. Along with the constant line of trains loaded with coal and coke that passed through the town, as many as eight passenger trains a day stopped in Appalachia. In the 1920s and 1930s the red brick depot was a busy place as mail, freight, coal, and passengers arrived and departed all hours of the day and night.

The population of Appalachia reached thirty-five hundred in the 1930s. The town provided entertainment, shopping, and services which were not available in the camps. On weekends, passenger trains and automobiles converged on the town, and Saturday nights were particularly crowded with families. The town, of course, also had its share of fights and disturbances as miners came to spend their weekly pay.[8]

The decline in the coal industry and the abandonment of nearby company towns took a heavy toll on Appalachia in the 1950s. As railroad service decreased and automobiles allowed people to travel farther away for recreation and shopping, the crowds of customers dwindled. Today Appalachia is considerably smaller than it was at the time of the coal boom. It now has less than half its peak population.[9]

Appalachia was an important link to the coal camps of Derby, Stonega, Roda, Osaka, Imboden, Exeter, and the railroad town of Andover, which are located off VA 78 and VA 160. Appalachia's depot, built in 1925, served the L & N, the Southern, and the Interstate. The last passenger trains ran in 1953, and the depot was closed in the late 1950s. At the west end of Appalachia on U.S. Business 23 is Bee Rock Tunnel, once featured in *Ripley's Believe It or Not* as the shortest railroad tunnel in the world, with a rock shaft of forty-seven feet. In August, the town hosts an annual Coal and Railroad Days Celebration.

Appalachia is the gateway to the Stonega mining district, a series of coal towns developed by the Stonega Coke and Coal Company. The largest is Stonega, located on VA 78 north of Appalachia. Developed by the Virginia Coal and Iron Company and expanded through its lessee the

Stonega Coke and Coal Company, Stonega became the center of a series of mining collieries developed north of Appalachia. Today Stonega and the series of small settlements surrounding it provide good examples of early-twentieth-century company towns and mine sites. Although rapidly disappearing, especially during the 1990s, there are still many structures remaining and hundreds of people living in homes built up to one hundred years ago.

Gen. John Daniel Imboden and Gen. Rufus Ayers brought the Virginia Coal and Iron Company to the coalfields around Big Stone Gap in the 1880s. By 1890 the company was developing mines along Callahan Creek, the future site of Stonega. Both Imboden and Ayers worked for Virginia Coal and Iron, which was founded in Connellsville, Pennsylvania, in 1882. They bought tens of thousands of acres of land in southwest Virginia and helped to arrange the merger of several area coal and iron companies into Virginia Coal and Iron. Ayers lived in Big Stone Gap and later served as director and vice president of Virginia Coal and Iron, the biggest developer of coal lands in Wise County by 1900.[10]

John K. Taggart, who had experience in coal mining, land leases and titles, coke production, engineering, and the designing of mining towns, took charge of the Wise County properties in 1890 and designed and supervised the first mining operation for the company. In June 1890 he arrived in the Callahan Creek area and immediately began surveying and mapping. By August he had built the first coke oven and was preparing samples of coal and coke for analysis. The railroads were still more than four miles away when Taggart laid out the town of Pioneer and surveyed mine sites. The coal and coke samples he forwarded to Pennsylvania were rated as superior, and the company stepped up its efforts to haul equipment into Pioneer. Meanwhile, Taggart planned for houses, a tipple, and rows of coke ovens.[11]

Several events brought the development of mining on Callahan Creek to an abrupt halt. The Panic of 1893 disrupted industry across the country. Several suits involving titles raised questions about Virginia Coal and Irons's rights to the land it was developing. The company also had problems negotiating an arrangement with the railroads to finish laying tracks to the mines. The Virginia Coal and Iron Company finally worked out an arrangement where it built its own spur up Callahan Creek and connected with the Louisville & Nashville Railroad. Construction resumed in 1895 and Taggart opened the first mine. By early 1896 there was a battery of coke ovens in operation, houses lined the area along the creek, the railroad reached the mines, and the name of the town was changed to Stonega.[12] Unfortunately, an explosion took the life of Taggart and another miner on 23 May 1896 as a new series of coke ovens was being built. The company sent a replacement superintendent from Connellsville and construction continued. By 1897 the town had more than one hundred houses and a store, offices, and company buildings. The ovens produced some of the best quality coke in southern Appalachia. Mining in the Stonega area began to boom.

The town was built to high standards and soon developed a reputation as a good place to live and work. Local journalists noted that housing at Stonega was "comfortable and convenient" and "much above the average of mining towns." By 1898 rail excursions were being run to Stonega to show visitors the marvels of life in a company town.[13] Virginia Coal and

Daniel B. Wentz served as the president of Stonega Coke and Coal from 1904 to 1926. From Edward J. Prescott's *Story of Virginia Coal and Iron Company* (1946).

Iron worked the Stonega mines until 1902, when it leased the lands to the Stonega Coke and Coal Company of New Jersey. Stonega Coke and Coal was a lessee of Virginia Coal and Iron, but it operated more like a subsidiary. The Wentz family of Pennsylvania, major stockholders in Virginia Coal and Iron, ran Stonega Coke and Coal. John S. Wentz served as president from 1902 to 1904 and was followed by his son Daniel B. Wentz, who served from 1904 to 1926. The Wentz family dominated the operations at Stonega and later moved the company headquarters to Big Stone Gap. The Wentzes understood the need to provide a good living and working environment for their employees and had the resources available to do so. Stonega Coke and Coal provided good housing, wages, and medical and social services and developed a good working rapport with its workers.[14]

Stonega Coal and Coke absorbed several smaller coal companies and in 1909 took over the nearby company towns of Keokee and Imboden. The company's settlements at nearby mining collieries developed into a series of small towns served by Stonega. Stonega Coke became the largest founder of company towns in Virginia, establishing a total of ten, and employing as much as 70 percent of southwest Virginia coal miners by World War II.[15] The town of Stonega was expanded by Stonega Coke after 1902, and there were more than 1,200 residents, 350 coke ovens, and 500 dwellings. Nearby, the company developed mines at Roda, Osaka, and Arno. At the same time, it expanded operations at Keokee and Imboden. The collieries of Dunbar, Derby, and Exeter were later added.[16]

Stonega Coke recruited laborers from the surrounding countryside, but the supply of workers was soon exhausted. Immigrants were brought in to work in the mines, giving the area an ethnically diverse population. Many of the workforce were Polish, Italian, Hungarian, and Slavic. African Americans from upper Alabama were also brought to the Stonega district. Virginia coal mining towns were segregated, with separate areas for white, immigrant, and black workers and their families. They attended different schools and churches and often lived in old, low-quality houses. At recreation centers, they were served in partitioned areas.[17] While Stonega Coke and Coal did not treat black and white residents equally, it did provide a better quality of life for blacks than was available in rural areas of the Deep South. By 1916 the population of Stonega had grown to well over 2,400 residents, 500 of whom were African American.

The First World War brought an expansion of coal production and Stonega's population continued to grow. The war cut off the supply of immigrants, and the new miners were almost all native whites and African Americans. Later efforts to recruit immigrant workers after the war were unsuccessful, and by the 1930s only a small portion of Stonega area miners were foreign born. The southern white population comprised about three-fourths of the workforce, another 20 percent was African American.[18]

Stonega employees received Christmas bonuses and charitable packages in times of need, often delivered personally by members of the Wentz family. A number of benevolent and relief clubs were formed with prompting and assistance from the company. Workers in the Stonega Coke and Coal mines were generally satisfied with the working conditions and the standard of living that the company provided. There appears to have been very little union activity in the area in the 1910s or 1920s when nearby communities in West Virginia saw open guerilla war break out between companies and unions. As a way of showing the company's commitment to its workers, Stonega Coke built new towns in the 1920s to even higher standards and expanded its services.[19]

Like many coal companies, Stonega Coke and Coal began to feel the effects of overproduction and falling coal prices in the 1920s. Near the end of the decade, the company was forced to cut wages and occasionally shut down. Operations were curtailed at Keokee and Osaka. The Great Depression of the 1930s made matters worse. Once a model of corporate paternalism, the Stonega Coke and Coal Company began to decline. Even so, there does not appear to have been any significant union activity in western Wise County until the 1930s. With New Deal legislation in place that recognized and legitimized unions, Stonega mines were the target of union organizers. Although the company had used firings and evictions to thwart union activities at times, they did not have to actively discourage workers from joining the union. And, although there were scattered incidents at Imboden and Derby, there was little violence. Stonega Coke and Coal, however, slowly gave in to the union. By 1935 there was a certain degree of cooperation between the union and Stonega, and the union began to assume many of the small social responsibilities for the community, such as planning holiday activities, overseeing relief funds, and conducting charity drives. In the 1940s the UMWA offered medical and retirement services and the company greatly reduced their social programs and services.[20]

Although World War II brought a brief resurgence in mining, Stonega Coke reduced its investments in the area, modernized existing operations, and sold its properties after the war. Mechanization had the biggest impact. By the late 1930s trackless loaders replaced hand loaders, and by 1950 many homes in Stonega and the company's surrounding towns were empty. In 1953 the company inventoried its properties and sold homes to individual miners for an average of $230 each. Other properties were sold or demolished. Stonega was no longer a company town.[21]

In 1964 Stonega Coke and Coal, which had once employed four thousand men in the area, absorbed the Westmoreland Coal Company and changed its name to Westmoreland Coal. Today the company employs fewer than one hundred men in the Appalachia area. Although mining has declined in recent years, Stonega and the surrounding towns still have hundreds of residents, and many remains of company buildings and structures are still intact. The landscape and layout of these towns reveal much of the area's coal mining past. All of these sites are within a few miles of Appalachia and can easily be reached from VA 78, VA 708, and U.S. Business 23. They vary in size, structural integrity, and population but are worth a visit since they reveal interesting aspects of the mining heritage of the Stonega district.

Andover

About two miles outside of Appalachia on VA 78 lies the railroad town of Andover (pop. 250), which served as Interstate's main railroad yard for the surrounding coal camps of the Stonega Coke and Coal Company. There were a series of disputes between the Virginia Coal and Iron Company and the Louisville & Nashville Railroad during the 1890s over the hauling of coal from the Stonega mine that led to the formation of a new railroad company, the Interstate Railroad, a subsidiary of Virginia Coal and Iron. The Interstate Railroad operated along a six-mile track from Stonega to Appalachia. In 1902 and 1903 it extended branch lines into Osaka and Roda.

Andover became an important rail center as the railroad expanded into nearby collieries. Railroad workers lived alongside the yard, in company houses built by Virginia Coal and Iron. The Andover yards were enlarged in 1921, with the addition of machine shops and a three-story office building completed in 1922. During the 1920s the Andover yard was the center of Interstate activity. The railroad had a fleet of more than three thousand coal hoppers and connected to four major railroads. The

Interstate Railroad operated until 1961, when it became part of the Southern Railroad system.

The Andover yards are now the site of the Norfolk-Southern Railroad's Andover Yard, part of the Pocahontas Division. The yard has seen little traffic since the late 1990s, and many of the facilities and service areas have been abandoned or dismantled. The houses along the tracks date from the 1910s and early 1920s. Just past Andover on VA 78 is the Central Supply Company Building, constructed in 1920 as a retail outlet for mining and mill supplies and general building materials. A series of tipples were still in use along VA 78 near Andover in the 1990s but have since been demolished.[22]

Derby

Derby (pop. 350) is located on VA 686 (Derby Road), off VA 78, just past Andover. Derby has a unique look among area coal towns in Wise County because red brick and hollow tile block were used to construct many of the buildings. Stonega Coke and Coal used these state-of-the-art materials to build the town in an effort reaffirm its commitment to its workers and prevent unions from gaining a foothold in their company towns.

In reaction to the West Virginia Mine War taking place across the Tug Fork, Stonega Coke and Coal wanted to show that their employees did not need the protection of a union. They believed that well-paid workers with nice homes would not be attracted to the union movement, so the company set out to create a showpiece community as a model of welfare capitalism.[23] Derby (which received its name because the company directors interrupted a trip to the Kentucky Derby to check the town's progress) operated four mines, a large tipple, and a loading facility. Additional preparation and loading facilities were added in the 1930s.[24]

Unfortunately, Derby became the site of one of the district's few mining tragedies. On 6 August 1934 a gas explosion at mine No. 8 killed seventeen miners, including the foreman, and injured many others. The death toll would have been much higher, perhaps more than one hundred, if the explosion would have taken place an hour later, when the entire crew would have been inside. The seventy-five men who survived had to walk through a long-abandoned section of the mine filled with water. With just a few inches of space between the water and the roof of the mine, the miners were barley able to keep from drowning. It was the worst mining disaster in the Stonega district and became the subject of a popular local folksong:

> Do let us all take warning, from Derby people's fate,
> And get ready for the Judgement before it is too late;
> Then if the mines in your town should happen to explode
> You'll be ready for the Judgement when you are called to go.[25]

The area was graded over after the explosion and the damaged houses were demolished.[26] The site can be found at the hump on VA 686 (Derby Road), about one-half mile from VA 78, where there are two railroad crossings.

The entrance to Derby is about one-quarter mile farther down the road. One can immediately see the unique design and coloring of the two-story, two-family foursquare houses with pyramidal roofs, full front porches, double entrances, brick chimneys, and coal houses. Another half mile up the road, on the left, is a supervisor's home, and on the right is the Derby United Methodist Church. The series of low-quality homes in this area were designated for African American miners and their families. Farther along on the left are the remains of two company office buildings and the foundation of the miners' bathhouse.

Osaka

Along VA 78 is VA 685 (Roda Road), which leads to Osaka (pop. 300). The Osaka colliery was built by the Stonega Coke and Coal Company in 1902 and was a very busy place in the early twentieth century. There were waiting lists for company housing and separate boardinghouses for single white and black workers.

Osaka was noted for its hundreds of beehive coke ovens that produced up to twenty-five loads of coke daily. Long rows of coke ovens between Osaka and Roda were usually manned by black workers, and part of the settlement became known as "Slabtown" because a series of older homes were moved onto concrete slabs for African American workers.[27]

The town prospered until 1927, when the mines closed, but many miners continued to live here while working in nearby mines, especially those at Roda. The mines at Osaka were opened once again in the 1950s, but the coke ovens were abandoned. By the early 1950s many of the company houses were empty. In 1953, Stonega Coal and Coke began to sell, and in some cases demolish, structures at Osaka. At that time, the company inventory showed a ten-room boardinghouse for black workers, barns and garages for employee horses and automobiles near the company store, and a small church for black residents.[28] The center of the coking operations were located where the Westmoreland supply complex was built in the 1960s.

A drive through Osaka reveals much of its coal mining past. Abandoned mining equipment, rows of uniform houses, and a winding railroad spur through the hollow reflect the typical layout of a company coal town. The mines at Osaka were operating again in the 1960s and sporadically during the 1980s, and equipment is stored in a series of brick buildings on the main road into town. About one-half mile farther is the site of the last coal tipple, destroyed in the 1990s. Just past the tipple, on the right, is the entrance to one of the mines.

The most prominent building in Osaka was the Osaka Company Store, built in 1925. The imposing two-story brick structure with large plate-glass windows was built alongside the railroad tracks facing the entrance to the town. It remained in use until the 1950s and was not demolished until 1999. The foundation lies to the right of the road, over-

The Osaka Store, built in 1925, dominated the entrance to the Osaka and Roda collieries. It was demolished in 1999.

grown by high grass. Just beyond the site of the store are about thirty houses and a prominent superintendent's home on a hillside.[29]

Roda

On VA 685, about one-half mile beyond the Osaka store site, is the town of Roda (pop. 200), named for the rhododendron bush common on the local hillsides. The colliery was developed in 1903 by Stonega Coke and Coal, where many Hungarian immigrants and African Americans were once employed.[30] Roda's residential area was called "Hunktown" because of the concentration of the two ethnic groups living there.

Five mines were opened at Roda from 1903 to 1955, and the town was home to hundreds of residents and a large company store. There were two large wooden tipples near the entrance to Roda, and the main line of the Interstate ran between the two. Electric mine cars brought coal from the mines, and night and day mine runs hauled loaded coal hoppers to Andover. The smaller of the two tipples was destroyed in the 1930s; the larger was demolished in 1982 and was replaced by a four-track loading site used by Westmoreland until the mid-1990s.[31]

Little remains of Roda today except for some scattered houses along VA 685. The former superintendent's home on the hillside to the left has been modified in recent years but was featured in a 1946 publication, which showed a series of wooden steps leading to a large two-story brick-and-frame hillside home with two brick chimneys and a deep wrap-around porch with white railings.[32]

The sites of the mines and tipples are overgrown, and the few remaining houses have seen better days. However, the Roda Baptist Mission Church, a small white wooden frame structure built in 1921, is still in use today and was the site of many revivals during the 1920s.[33]

Roda was also the home of Elizabeth Collins, a legendary woman who served as a combination minister, doctor, counselor, and friend to thousands of people living in the coal camps and mountain communities around Wise County. Collins, known as Toddy, arrived in the southwest Virginia mountains in 1903 and spent the next forty-three years serving the area's remote communities. As a young girl, she attended a private Episcopalian school in Richmond and became interested in mission work. When she read about the impoverished communities of southern Appalachia, she decided to dedicate her life to helping the people of the

Above: This supervisor's home is one of the few remaining in Roda and has undergone many renovations since it was constructed in the early twentieth century.

Right: The Roda Baptist Mission Church, another of the few remaining buildings in Roda, was built in 1921.

mountains. When she arrived in Wise County, the Stonega Coke and Coal Company was in the midst of developing coal towns along Callahan Creek. Although many of the local residents were suspicious of her at first, the company realized her potential usefulness to miners and their families. Stonega Coke and Coal offered her a company cottage in Roda, where she soon held church services, literacy classes, and organized domestic projects for miners' wives. Over the next few years she became acquainted with thousands of people in the coal camps and hollows of the area and helped organize church services and projects at Osaka for the African American community. She served as a nurse for the company doctor and delivered babies, tended to the sick, made coffins, and held funeral services for the deceased. At her insistence, the company built a small school and nondenominational church at the head of Roda hollow. She became known as the Angel of Happy Hollow (a nickname for Roda), and her reputation spread throughout the country. So many people were touched by her work that when she died in 1948 the mines were closed and hundreds accompanied her funeral procession. *The Saturday Evening Post* ran a feature article about her life and works.[34]

Stonega

The center of mining in the Stonega district was Stonega (pop. 300), located on VA 78 about three miles north of Appalachia. The community was originally known as Pioneer and was the first coal town established in the Callahan Creek area of Wise County. Built in the 1890s, the busy coal mining community attracted thousands of miners and their families in the early twentieth century. The town gained Stonega Coke and Coal a reputation for offering high-quality housing and varied facilities. The company operated a slaughterhouse, bakery, and boardinghouses and built churches, company stores, schools, and leisure facilities in Stonega and its many collieries. There were also company baseball teams, brass bands, and tennis courts. At its peak, Stonega had 702 houses, full medical and entertainment facilities, and a large commissary with refreshment parlors, barbershops, and amusements. By 1916 there was a full sewage system and fire hydrants.[35] A fire and police department oversaw activities in the town, and a small company hospital staffed by physicians and nurses attended to family health needs. When first established in 1902, the Stonega Hospital was the best medical facility between Louisville,

Kentucky, and Asheville, North Carolina. The staff handled medical emergencies and mining accidents, had first aid teams, and provided family health care and vaccinations. The hospital served the town until 1957.[36]

Just past Roda Road, VA 78 becomes VA 600, or Stonega Road. It crosses the railroad tracks twice and becomes the main street through Stonega. The area after the second crossing is lower Stonega, a series of four-room, single-story weatherboard houses with side-gabled roofs and front porches built for white miners and their families. When originally constructed, each home was surrounded by a white picket fence.

Three-quarters of a mile past the railroad tracks is central Stonega, where company offices, churches, stores, and recreational facilities were once located, and on the left is a row of garages for residents' automobiles. The main company buildings have been razed, but the Stonega Baptist Church and several brick homes are on the left, with large landscaped lots. The best houses, built for management, were located on a hillside in the central area of town, along Park Place Road. Beyond these are a section of two-story houses, known as "Quality Row," reserved for managers and skilled workers.[37]

Less than one-quarter mile past this area is a road on the right. It leads to a large brick building that was once a machine shop. Just past the

Top: Miners' homes in Lower Stonega, Virginia, circa 1915–20. These homes are still occupied today along VA 600. Courtesy of Hagley Museum and Library, Westmoreland Collection.

Left: The large building on the left has served as schoolhouse, store, and community center in Upper Stonega. This photo was taken around 1920. Courtesy of Hagley Museum and Library, Westmoreland Collection.

Hundreds of coke ovens lined the railroad tracks in the Stonega mining district, circa 1920. Courtesy of Hagley Museum and Library, Westmoreland Collection.

machine shop is a bathhouse, built after 1916 when a company medical staff report insisted that coal dust was unhealthy for miners and their families. The report suggested separate facilities where miners could change clothes and bathe after a shift in the mines. This area was also the site of hundreds of Stonega coke ovens that spread out along the tracks from the mines. All that remains of the Stonega coke ovens is a pile of collapsed stones in the brush between the machine shop and the bathhouse.

Continuing on Stonega Road another quarter mile, there is a large two-story white wooden structure with an elevated porch, large window bays that have been sealed, and a black bell mounted above the door. This building served at various times as a school, store, recreation center, and church. Just past the school are older, lower-quality two-story, two-family houses that were used for African American workers and their families. They did not have electricity and were some of the last homes to be upgraded. The area still has a small community of African Americans and a small one-story wooden frame church with a square bell tower. These few homes are the upper limit of Stonega today as mining operations continue to move closer to the town. Perhaps another fifty houses were along this road at one time.[38] Beyond Stonega was the site of the Osaka No. 2 mine and a lumber mill. A large preparation plant and loading facility are located farther along the road at Wentz.

Dunbar

Dunbar (pop. 130) is located on VA 603 (Dunbar Road), off VA 78 north of Appalachia. It was built in 1918 on Pot Creek, just above the town of Roaring Fork, to mine two newly identified coal seams, the Taggart and Marker. Laborers cleared the forests and laid the foundations for a tipple and railroad line. A trough conveyor was built above the rail line, and coal was lowered to a tipple below. Carpenters constructed some single-family dwellings but soon realized there was not enough room to build the 250 houses planned by Stonega Coke and Coal engineers. Instead, the plans were changed and they built ten-room double-occupancy houses to accommodate two families. They were wood frame dwellings with poplar siding and brick foundations. Inside they were wired for electricity and had plastered walls and open fireplaces. About one hundred homes were eventually built, and the mines were in operation by 1919.[39]

From Roaring Fork, the road on the left leads to Dunbar, about four and-one-half miles from VA 78. Five mines operated at Dunbar from 1919 to 1946. The layout of Dunbar is typical of many mining towns described in this guidebook: a uniform style of houses along a railroad spur and a community church. At the entrance to town is the Dunbar United Methodist Church, built in the 1920s, and directly opposite is the remaining foundation of the company store. Next to the store was the local recreation center and theater. Dunbar's double-occupancy houses were tightly packed together in two long rows on the flatland along the railroad tracks. They were two-story, side-gabled frame dwellings with front and back porches and small yards. About thirty-five homes are still occupied.

About one and one-half mile farther north on VA 603 is the site of Pardee, a coal town demolished in the 1980s for current mining operations. The Blackwood Coal and Coke Company built the town in 1903 and named it after its president, Calvin Pardee. The mines produced an extremely high-grade bituminous coal that was marked with a thin stripe of red paint as it was loaded. After the original company store burned in 1940, a large new brick store, or commissary, was built to provide miners

The center of Dunbar in the 1920s was the site of the community church, company store, recreation center, and theater. From Edward J. Prescott's *Story of Virginia Coal and Iron Company* (1946).

with equipment and supplies, as well as a variety of consumer goods for local residents. This store was featured in the 1981 film *Coal Miner's Daughter,* and parts of the town were restored for several scenes in the film. The town's destruction allowed for a new mining center to be built in the 1990s with a loading facility and preparation plant.[40]

Pine Branch Coke Ovens

From Roaring Fork a rough road to the right leads north along the railroad tracks. The pavement ends after a few miles, and a graded road continues to the site of Pine Branch's coke ovens. In 1946, when the Pine Branch mines opened, the coke was prepared, crushed, and loaded into railroad cars at the site. In the 1950s, the company filled about fifteen to twenty carloads a day, using horse- and mule-drawn wagons to transport coal from nearby mines to the ovens. The operations were later mechanized, and in the 1960s a conveyor system was added that carried the coke to a crushing machine and loaded it directly into coal hoppers

The Pine Branch coke ovens were operated until the late 1960s.

on tracks that ran along both sides of the ovens. The Pine Branch coke ovens remained in use until the late 1960s.

Today the road to the site is busy with a steady flow of coal trucks from nearby mining operations. The coke ovens, silent now for more than thirty years, are an imposing sight. They lie along one side of the road, stretching over an area of approximately three hundred feet. About 60 of the original 180 ovens remain, some with their large metal doors intact.

West on U.S. 119 from Letcher County

Travelers heading west on U.S. 119 from Letcher County have the opportunity to visit a number of interesting coal mining sites. Harlan County (pop. 33,202), just west of Letcher County, is one of east Kentucky's most famous counties and the site of several major events in the struggle to unionize coal miners. It was created in 1816 from part of Knox County and was named for Silas Harlan, a hero of the Battle of Blue Licks.

During the nineteenth century most residents survived by farming and bartering, but by the 1880s logging developed as a thriving industry on the Cumberland River.[41] Harlan County developed slowly, and it was not until the arrival of the Wasioto & Black Mountain Railroad in 1911 that its resources could be easily exploited. The coal boom of the early twentieth century changed Harlan County quickly, and production and population grew at an explosive rate. With fewer than 10,000 residents in 1910, the county's population leaped to 31,000 in 1920 and 64,000 by 1930. By the 1920s Harlan County was booming with record production, and many company coal towns. The largest employers in the county, including International Harvester, U.S. Steel, and Peabody Coal, remained staunchly anti-union and harnessed the power of county government, the local media, and deputy sheriffs to prevent organization. During the 1920s, union membership remained low and union organizers were closely monitored and harassed.[42]

With Harlan County hit hard by the Great Depression of the 1930s, unions seized an opportunity to increase their membership rolls. There was great opposition by coal operators most of who refused to recognize any of the protections provided under New Deal legislation. The coal companies used every means available to intimidate and harass union organizers and sympathetic miners. In the 1930s many of these conflicts became violent. Company guards assaulted union men, and public demonstrations were disrupted. Union men retaliated by dynamiting mines and attacking company men. At one point in 1931 there was a murder almost daily, drawing national attention. The press labeled the area "bloody Harlan County." Violence continued and the county became a testing ground for Roosevelt's New Deal legislation. Harlan County was also one of the few places where violence broke out between rival unions.

By the late 1930s, the UMWA temporarily prevailed over many coal companies and local employees' associations. But by the 1940s, changes within the coal industry resulted in a need for fewer miners. From 1940 to 1960 the number of miners employed in the county plummeted from 15,800 to 2,200. That number dropped further with the closing of the area's two largest mining operation in the 1960s at Benham and Lynch.[43]

Today, along with many former coal towns within its borders, Harlan County offers visitors a variety of sites. Kingdom Come State Park offers terrific natural scenery and outdoor recreation. Two festivals, the Harlan County Homecoming and the Poke Sallet, both held in June, provide great entertainment. But Harlan County's newest attractions are its development of heritage tourism sites that focus on the coal industry. Benham and Lynch, two company towns that are undergoing great restoration, offer the visitor a rare opportunity to see well-preserved coal towns and visit mining sites.[44]

Harlan

Harlan (pop. 2,081) is located on U.S. 119 and is the seat of Harlan County and a regional government and commercial center. Harlan was originally called Mt. Pleasant when settled in 1819 but was formally incorporated in 1912 as Harlan to honor Maj. Silas Harlan, who fought in the American Revolution with George Rogers Clark. The town prospered with the rapid expansion of the Harlan County coalfields in the early twentieth century. The downtown commercial district, south of Central Street, features many structures dating from Harlan's boom period of the 1910s and 1920s. On a typical Saturday hundreds of miners and their families frequented the shops, businesses, and theaters along these streets. There are rows of solid brick one- and two-story buildings, often with large display windows, typical of early-twentieth-century commercial architecture in the region. The heart of the district, between Central and Clover Streets, has been added to the National Register of Historic Places.[45] The Harlan County Courthouse, a classical revival, brick-constructed, limestone-veneered building with a jail annex was often the site of legal battles over unionization in the 1930s.

The town has recently been rescued from the many floods that have plagued its residents over the years. A $31 million flood control project reversed the flow of the Cumberland River's Clover Fork and diverted its waters into a series of tunnels leading to the outskirts of the town.[46]

The Harlan Appalachian Regional Hospital is the second largest of a series of ten regional health care facilities in the greater Big Sandy area. It was built in 1956 as part of the Miners Memorial Hospital Association, a UMWA project that brought health care to southern Appalachia after the major coal companies sold their company towns and curtailed medical services. As president of the UMWA, John L. Lewis hoped that the union could continue offering health care to miners, but dwindling membership and depressed coal prices jeopardized the project. By the early 1960s the project collapsed. In 1963 a group of religious organizations developed a plan to save the system. With federal and state monies they reorganized and opened the hospitals as Appalachian Regional Hospitals. The eleven union hospitals became part of Appalachian Regional Hospitals operating throughout the southern Appalachian coalfields. In 1986 they were reorganized as Appalachian Regional Healthcare and continue to operate throughout the mountains today.

In 1991 a monument was erected to honor the many people killed in mining over the years. Located alongside the Harlan County Courthouse, the black and gray granite monument bears the names of the 1,282 miners who perished in Harlan County mines. Beginning with 1912, the monument lists the names of miners killed each year. In 1927, the worst year, fifty-five lives were lost. Early entries such as Bozo Radulovich, Matteo Giviriccin, and Augustine Antonio show the various ethnic backgrounds of people who worked in the mines. Family tragedies are also evident, as in the 1932 accident that took the lives of six Masingale brothers. The monument has room for more than one hundred additional names and is dedicated to the hundreds of coal miners who "sacrificed their lives while supporting a family and a nation."[47] A good time to visit Harlan is during

A monument erected in downtown Harlan in 1991 honors the many workers who lost their lives in Harlan County mines.

its Heritage Days Festival held on Labor Day weekend each year. Activities celebrate the pioneer and mining heritage of the region.

Evarts

Evarts (pop. 1,101) is a quiet community located off KY 38 about eight miles from Harlan, at the confluence of Clover Fork and Yocum Creek. The road enters the town's former commercial section. Buildings on the main street were once active stores and businesses. Just past the town is the former L & N Railroad line and some mining equipment. A few houses from the early 1920s are south of town.

Evarts was incorporated in 1855 and named for a local farm family. It was later developed by the Peabody Coal Company in 1919 and gained a reputation for union activism in the 1930s.[48] The site of a series of violent incidents during the struggle to unionize "bloody Harlan County," the town is best known for the 1931 Battle of Evarts. On the morning of 5 May, a procession of two hundred armed miners poured into the town following a series of evictions of union men and their families. As a convoy of Black Mountain Coal employees transported the possessions of nonunion miners to a new coal camp, they passed near Evarts, where they were ambushed. Although it is not clear how many people were involved in the attack, more than a thousand shots were fired. Several company men were gunned down as they attempted to flee or surrender. When the attack ended, three company men and one miner were dead; four wounded company men fled the scene. The event gained national attention as part of the county's continuing labor crisis and led to the arrest and conviction of organizers as conspirators and murderers, which in turn provoked a full-scale strike, bringing radicals and communist organizers into the region.[49]

From Evarts, adventurous travelers can continue on KY 38 to KY 179 as an alternate route to Cumberland, Benham, and Lynch. This is a fairly grueling drive across Black Mountain with some sharp turns and steeps inclines.

Cumberland

Cumberland (pop. 2,611) is located on KY 2179 off U.S. 119 and is the largest town in Harlan County. Settled in the 1870s, it was once known as Poor Fork. As railroads opened the mining areas of Benham and Lynch in the early twentieth century, Poor Fork began to grow as a commercial center. The extension of U.S. 119 in the 1920s connected some of the larger settlements in Harlan County, and the city was renamed Cumberland, which local businessmen considered to be a more respectable name for their community.

As an independent town, Cumberland was not under the strict control of the coal companies; however, it was greatly influenced by them, and local businesses realized they owed their existence to the coal economy.[50] The town's once-thriving commercial district, listed on the National Register of Historic Places, had theaters, shops, lodges, and restaurants. It was a popular place to shop, do business, get involved in politics, or simply have a good time. West Main Street, located near a bend in Poor Fork, was the heart of Cumberland's business district. Still standing are a series of one- and two-story brick buildings typical of those in regional commercial centers.

In the 1980s Prof. James Goode, of Cumberland's Southeast Community College, began organizing an archive for materials relating to the coal mining heritage of the area. The archive contains periodicals,

Cumberland's commercial district, placed on the National Register of Historic Places in 1995.

newspaper clippings, company records, photographs, and local histories. In 1985 Arch of Kentucky donated more than four thousand photographs and records to the archive. Researchers interested in these materials should contact the college.[51]

Benham

Benham (pop. 599) is located on KY 160 east of Cumberland. Built in the 1910s, it is one of the best-preserved coal towns in Kentucky. The town's recent efforts have focused on the Kentucky Coal Mining Museum, a project sponsored by the Tri-City Chamber of Commerce, the Kentucky Heritage Council, and the Appalachian Regional Commission. The museum now serves as a base for the further revitalization of the town, and many original company buildings have now been restored. The entire center of town has been designated as a historic district and is listed on the National Register of Historic Places.

Once an isolated settlement at the base of Black Mountain, with few residents other than occasional hunters and trappers, Benham became a coal town soon after surveys of the Cumberland Mountains revealed rich coal deposits.[52] Around 1900, International Harvester's mining division began leasing lands in Harlan County to extract a high-quality coal for its steel producing subsidiary, Wisconsin Steel. Through the use of broad-form deeds, International Harvester leased land in the Cloverlick Creek area, the present site of Southeast Community College in Cumberland. Although the nearest railroad was more than thirty miles away, International Harvester began to prepare the area for the large-scale mining of high-quality bituminous coal.

Engineers from the company entered the Black Mountain area and began laying out the mining operation in 1910, hauling heavy equipment and supplies by packhorses and mules. They quickly built the basic structures needed for mining, such as a tipple and bunkhouses, and a sawmill provided timber. In July 1911 large-scale construction began as the Louisville & Nashville Railroad reached the town. By 1912 the original portion was completed between the Scotts branch and Maggard branch of Looney Creek. Construction was rapid and the coal town appeared almost overnight.[53]

International Harvester recruited more workers as their mining operations expanded in the 1910s. The earliest miners were local mountain residents, but the mining boom throughout Appalachia quickly depleted this source of labor. Hungarians, Italians, Slavs, Czechs, Germans, Scots, and Irishmen soon provided the bulk of the workforce, and southern blacks, about one-quarter of the workforce, were recruited from Alabama, Georgia, and Virginia. Along with the expansion and diversity of the workforce came three churches: one for Protestants, one for Catholics, and one for African Americans. Two YMCA buildings were constructed to serve the recreational and community needs of the white and black residents, and a school tax was levied to hire teachers.

The outbreak of World War I in 1914 created an unprecedented demand for coal, and the operations at Benham and nearby Lynch were quickly expanded. The lower end of town was built by 1918 and became known as New Benham. By then the company had 520 houses and leased them to miners and their families. Although there was more variety in Benham, the basic four-room cottage was most common. These wood frame houses had two bedrooms, a kitchen, and a living room. Two-story duplexes, in which one family occupied each side, had plastered walls, an outdoor toilet, and a front porch. Additions were made to the basic design to create larger homes for managers, supervisors, and company officials. By the 1920s, indoor plumbing was added. Until that time, women used galvanized steel tubs to haul water to their homes from outdoor hydrants.[54]

All of the town's houses were constructed by 1919, and because mining at Benham continued to expand for the next ten years, there was a constant shortage of homes and a waiting list for company houses. There were few complaints about housing in Benham. International Harvester provided numerous services for miners, and company houses were well maintained. Painting and carpentry crews regularly serviced the homes, tenants were expected to keep their yards clean, and the company hauled away all trash and refuse.[55] Benham also had good quality medical services, with health insurance paid partly by International Harvester. In 1921 a hospital was built in the center of town and was staffed with physicians and nurses. Although the Benham mines were safe by industry standards, many health problems plagued the community as mining runoff and assorted contaminants spoiled the water. In addition, Benham's valley location made pollution from coal burning stoves and steam engines a constant source of irritation.

International Harvester was proud of its operations at Benham and provided a full range of services and facilities for its employees. A large commissary, or company store, was built in the center of town, along with a meat market, recreation center, and theater. A football stadium was later constructed, and local teams battled in the annual Coal Bowl. To maintain order, International Harvester had a private security force with much broader powers and responsibilities than a typical mountain police force. They recorded any unusual behavior that might threaten production and closely monitored any suspected union activities that might disrupt production.[56]

By the 1930s Benham had 4,500 residents. Although miners worked together in the mines, Benham was segregated and had four distinct neighborhoods. African Americans, who made up one-quarter of the town's population, lived in Smokey Row and had separate social, religious, and recreational activities. Immigrant workers and their families lived in New Benham. Whites and middle management resided in the middle section of Benham, from Poplar Street to Hemlock Street. Company officials and supervisors lived in the fanciest section of Benham, known (as in many mining towns) as Silk Stocking Row. The cemetery, at the extreme west end of Benham, maintains these same ethnic and racial divisions.[57]

The drift mines at Benham eventually reached five miles into the mountain, with four mines along the Benham spur and one on Looney Ridge. The total output of these five mines was more than thirty million tons of coal by the time they closed in 1963. Working conditions were generally good and International Harvester offered extensive safety and medical training. An accident insurance program covered all miners and paid half wages for as long as two years for anyone injured on the job. In the event of death, families continued to receive wages for three years. These benefits were quite generous compared to other coal companies.[58]

Despite the efforts of International Harvester to take care of its employees, the Great Depression took its toll on the company and the town. New Deal legislation gave a new boost to union activities in Harlan County, and the Progressive Mine Workers of America (PMWA), affiliated with the American Federation of Labor, organized Benham in 1932. It was the first and only union to represent the miners at Benham. Although the UMWA made many efforts in the 1930s and early 1940s to bring Benham miners into their ranks, these efforts failed and occasionally led to violence. In the summer of 1941, crowds of UMWA supporters swarmed into Benham from nearby Lynch. Dynamite charges were set, property was destroyed, and snipers fired into the town. In separate incidents, two

African American miners were killed. At one point a rowdy crowd of three thousand men descended on the town. After intervention by the governor and the state police, the crowds were contained and a new election was held. The UMWA was defeated and temporarily gave up its efforts in Benham. They later returned but continued to be defeated in every election held in Benham until the 1980s when the mines finally closed.[59]

Although the PMWA triumphed, Benham was already in decline in the 1940s. The mines had employed twelve hundred people in the 1920s. By 1929 they employed about nine hundred, and the Great Depression reduced the number to three hundred. The 1940s brought a brief boom, but modernization reduced the need for manpower to about five hundred employees by the 1950s. In 1960 International Harvester employed fewer than four hundred people in Benham and no longer needed to maintain a corporate community. The company concentrated its efforts on mining coal, not operating a company town, and began selling houses and properties to employees and private businesses. In 1961 Benham was formally incorporated and ceased to be a company town. The completion of new highways in the area ended much of Benham's isolation, and the company store closed in 1963. The mining areas, facilities, and offices were maintained by International Harvester until the 1970s when the Scotia Company purchased these properties. They continued mining in the area, using modern technology and further reducing the workforce. Arch of Kentucky purchased Scotia in the mid-1980s, and mining in the Benham area was abandoned.[60]

Between 1920 and 1925 International Harvester created a central park area in Benham surrounded by company buildings that reflected the commercial building styles of the period. Most were located along the main highway, with access to the railroad in the rear. The largest was the Benham commissary. A three-story rectangular concrete and masonry structure with a flat roof and front parapet facing the park, it had two main entrances and large plate-glass display windows showcasing merchandise. The commissary was built on the site of the original company store, a wooden structure lost to fire in the late 1910s, and served area residents until the 1960s. By the 1980s it had suffered years of neglect and was being used for storage. In 1990 the Tri-City Chamber of Commerce purchased the building and began to collect artifacts for exhibition. State and federal agencies provided extensive funds for renovation, and under the leadership of Mrs. Bobbie Gothard it was opened in May 1994 as the

Kentucky Coal Mining Museum. The museum provides a complete picture of life in a coal mining community with exhibits that show mining equipment, a company hospital, items from the company store, and the interior of a miner's home. There is a video presentation on the history of the Benham area, a scale model of a coal tipple, and many photographs displayed throughout. Country music legend Loretta Lynn, known as the "Coal Miner's Daughter," has donated a special collection of artifacts displayed on the third floor. Outside there is a two-ton block of coal cut especially for the museum and mining machinery, including an electric locomotive that took men in and out of the mines. The sites of the main mines and the Benham tipple are directly behind the museum.[61]

Surrounding the park are eight other major company buildings. Directly west of the museum is the Benham Meat Market, a rectangular brick building where dairy products, meats, and produce were once sold. The market building had a basement that served as the town's ice house. Across the street, facing the park, is the former Benham Theater, a two-story brick building that was used as an entertainment and meeting center. East of the museum is the Benham City Hall, a one-and-one-half-story brick structure with a double-door entryway and large windows on the left side that once functioned as the company pay windows. The building was originally used for the offices of Wisconsin Steel. Next door is a two-and-one-half-story brick building that still operates as Benham's post office. Across the street is an irregular bungalow-style structure with a front porch and gable roof porch. It served as the Benham Hospital when completed in 1922.[62] The Benham Fire Hall was once housed in the nearby one-story frame structure. The Gothic-style frame and stucco Benham Methodist Church has been in use since its completion in 1928. It has a square bell tower and Tudor-style windows with stained glass.

Across from the Kentucky Coal Mining Museum, on the hill above the park, is the Benham School, a two-story brick building with a beautiful offset tower and a steeple sheltering the main entrance. Built in 1926, it served as the town's elementary and high school until 1961; it continued as an elementary school until 1992. In 1993 the property was thoroughly renovated and became the Benham Schoolhouse Inn, opening in 1994. Thirty classrooms have been turned into lovely guestrooms and are filled with antiques and cherry wood furnishings. Much of the interior of the building remains the same, including long rows of student lockers. The gymnasium is now a restaurant and banquet hall. The Benham

The Benham Post Office has been in operation since the 1920s, and Benham City Hall once served as the offices of Wisconsin Steel.

Schoolhouse Inn is managed by the Southeast Education Foundation, a nonprofit organization that supports programs at Southeast Community College. The college offers internships at the inn as part of its Hospitality Management Program. The Benham Schoolhouse Inn offers the warm atmosphere of a mountain inn and has become a favorite with locals as well as tourists.[63]

Company houses line the streets above the school, and the Benham area retains the look of an early-twentieth-century company town. A cemetery at the west end of town near Beech Street features many interesting tombstones that show the variety of workers at the Benham mines and a segregated section for black miners and their families. Visitors should make the museum their first destination to get further information. From Benham, it is a short drive to Lynch, where the mining district is the main attraction.

The original Benham School building was used as an elementary and high school until 1961. It was recently reopened as an inn.

The Kentucky Coal Mining Museum features exhibits that portray the lives of miners and their families.

Located east along KY 160, Lynch (pop. 900) is another well-preserved early-twentieth-century mining town with a great deal of mining equipment, company buildings, and houses still intact. Unlike people in many neighboring towns, local residents were able to save many structures from demolition and have made progress in developing tourism. Lynch appears to be a town suspended in time and is becoming a favored destination of many travelers.

Lynch was designed, constructed, and operated by the nation's largest corporation in the 1910s, J.P. Morgan's U.S. Steel. In 1917, at the height of wartime production, U.S. Coal and Coke, an affiliate of U.S. Steel, sent an engineer, L. A. Billips, to east Kentucky to locate a coalfield suitable for immediate development. He found a tract in Harlan County and arranged the purchase of fourteen thousand acres. U.S. Steel began building a railroad spur from Benham to mine sites three miles up Looney Creek. The coal was to be used in U.S. Steel's massive mills in Gary, Indiana.[64]

U.S. Steel built a model company town at Lynch that exceeded the standard of living established by the Consolidation Coal Company in Jenkins. Because engineers estimated that the area would yield coal for nearly one hundred years, they recommended that structures at Lynch be made of durable materials made to last well into the twenty-first century. U.S. Steel's architects designed a corporate community with quality housing, recreation facilities, and services.

The first residents of Lynch were Italian and Hungarian stoneworkers who lived in sparse temporary shacks while the town was constructed. They built stone foundations for the large corporate structures, the hospitals, and the offices. As houses were completed and mines began to operate, the population of Lynch grew dramatically, becoming a cosmopolitan community with German, Russian, Polish, Greek, and Croatian laborers. African Americans, escaping the dilemma of sharecropping and increased racial tensions, were recruited from the Deep South. Although U.S. Steel sought workers from nearby mountain communities, fewer than half the residents were southern whites.[65]

Only the highest quality materials were used to build rows of company houses with asbestos slate roofs, plastered interior walls, electrical wiring, and full plumbing. Four hundred duplexes and two hundred single-family cottages were constructed. The most common design was a

two-story, double four-room house with a full front porch with two separate entrances, a downstairs living room, closets, a kitchen, and bedrooms
upstairs. Small single-family, three-room houses were also constructed
where space was limited. These frame houses had solid stone wall foundations and roofs with asphalt shingles. Lynch planners tried to avoid the
monotonous appearance of many mining towns by varying exterior shapes
and details while leaving the basic interior design the same. The company
also painted the homes different colors, with yellow, white, gray, green,
and red being the most common. On the hillsides were cooler, more spacious residences for managers and engineers. These homes had as many as
eight rooms and comfortable porches and patios. By the 1920s U.S. Steel
had built more than one thousand houses in Lynch.[66]

Town planners created a pleasant suburban atmosphere in Lynch
with paved streets and sidewalks, churches, an independent school system, and parks for recreation and sporting events. One of the town's
showpieces was the Lynch Hotel, a long, three-story stone and wood
building with a circular portico and large white columns, that once stood
on the hillside above the town. The hotel was originally intended to
house single workers as well as visitors, but so many people moved to
Lynch that it was soon replaced by five immense boardinghouses. The
hotel became a luxurious place where visiting official and politicians
were entertained. It featured 130 carpeted, steam-heated rooms, shops,
restaurants, bowling alleys, pool rooms, a large dining area, and a theater.
Lynch hosted famous internationally known opera stars, jazz bands, and
radio personalities. World-famous Yugoslav tenor Tina Pattiera and
Metropolitan Opera's Italo Picchi performed here in the 1920s.[67]

The Lynch mining operation was the most modern in the world, and
its steel and cement tipple was the largest ever built. The mines were the
first to be totally electrified, with all sections fully lit, and they were also
the most productive. On 12 February 1923 the Lynch mine broke the
world record for a single day's output: 1,050 underground workers mined
7,089 tons of coal.[68]

By 1920 Lynch had more than seven thousand residents, and U.S.
Steel continued to acquire land in Harlan County. The Lynch district of
U.S. Steel soon reached along the Cumberland Plateau into neighboring
Virginia, covering an area of more than thirty-nine thousand acres,
including the entire area bordered by Poor Fork and Clovis Fork of the
Cumberland River. By 1930 the population was almost ten thousand. The

payroll of U.S. Steel at Lynch reached more than a quarter of a million dollars. When the payroll train arrived, both ends of the town were sealed off by company police and no one was allowed to leave the town until all employees had received their pay. By the 1940s the Lynch district employed more than forty-five hundred people.[69]

The town was named for Thomas Lynch, who headed coal and coke operations for U.S. Steel after 1900. Lynch helped shape U.S. Steel's corporate paternal attitudes and placed working conditions, safety, and the miners' standard of living high on the company's list of priorities. He developed strict codes for U.S. Steel mines and incentives for workers to put safety first. Rewards and citations were given to supervisors who oversaw the safest sections, and the motto "Safety the First Consideration" was placed in work areas and at mine entrances. Lynch died in 1914, but U.S. Steel named the new town for him, hoping that it would be the embodiment of his ideas, and during the 1920s many of these goals were met. Miners worked in a relatively safe environment with modern equipment, good pay, and benefits. In 1925 the daily minimum wage at Lynch was more than six dollars for a nine-hour shift, better than most other mines in the area. There was one tragic event at the No. 30 mine in October 1931 when a gas pocket explosion killed three workers and badly burned a fourth.[70]

Lynch was a booming community during the 1920s, but the Great Depression created problems for U.S. Steel and its workers. The cost of coal dropped and workers' wages were reduced. The company also curtailed services and brought in new managers who focused on cutting costs. During the 1930s Lynch experienced problems similar to other communities in the area. Violence, intimidation by company guards, and harassment of labor leaders was common throughout "bloody Harlan County" at the time. The worst disturbances were in 1934 when seven hundred employees were discharged, hundreds of other resigned, and those who resisted were driven out of town. When New Deal legislation encouraged union activities in the county, the UMWA infiltrated Lynch and organized its miners. Unfortunately, there was a great deal of violence in the area during the 1930s, much of it between competing unions. The Lynch UMWA, affiliated with the Congress of Industrial Organizations, fought the Benham chapter of the Progressive Mine Workers of America, who were affiliated with the American Federation of Labor. These clashes escalated in the late 1930s and peaked in 1941 when state and federal intervention finally brought an end to the violence.[71]

World War II created a great demand for coal, and Lynch saw its peak population during the war years. After the war U.S. Steel began to concentrate on modernization. The Lynch mines operated for more than forty years, but the need for manpower declined by the 1950s. With mechanization, the Lynch mines yielded more than one million tons of coal a year in the 1960s. By 1982, the Lynch District produced more than two million tons of clean, high-quality coal but with fewer and fewer employees.[72]

Lynch's coal reserves did not last one hundred years. By the late 1950s mining operations moved to nearby sites and Lynch began its decline. With fewer employees and less activities at Lynch, there was little need to maintain a company town. In 1963 U.S. Steel ended operations at Lynch and sold its properties to individuals and businesses. Lynch was incorporated and ceased to be a company town, although the town's machine shops and mining services continued to support coal production.

U.S. Steel further modernized the Lynch mines with continuous miners and longwalling methods. The main mines, portals 30 and 31, were exhausted by the 1960s, but nearby operations continued until the late 1980s. Coal was removed and sent to the Corbin Preparation Plant. Coal from Lynch saw domestic use by U.S. Steel, and was also shipped to Mobile, Alabama, for overseas operations. The main mines were sealed in 1963 but were worked again with newer methods in 1968. A conveyor system was built to haul coal from the top of the mine to the tipple. By 1983 the entire Lynch District employed only 960 workers. Strip-mining removed the last coal from Lynch, and operations ceased in 1991. Arch Mineral Corporation worked some of the areas around Benham and Lynch and had 400 employees in the late 1990s but closed its Harlan County operations in 1998. Today Lynch has about 1,150 residents, many of whom are retired miners and their families.

Despite the end of coal mining, Lynch is experiencing a resurgence as a tourist destination. The Cumberland Tourism Commission and the Kentucky Coal Mining Museum at Benham have been instrumental in preserving the area, and in recent years it has attracted thousands of visitors. A group of local Lynch residents, under the leadership of the late James R. Stewart, have been very successful in gathering artifacts and organizing reunions and are currently writing an extensive history of the region. Although many homes have been lost, much of Lynch is intact and has not been developed or modernized, making it appear much as it did fifty years ago. A walking tour brochure and map, available from the

Tri-City Chamber of Commerce and the Kentucky Coal Mining Museum, guides visitors through the town.

The entrance to Lynch, one mile east of Benham, is located on KY 160. Lower Lynch is the old African American community where many black families continue to live. Houses, both single and duplex design, line several side streets paralleling the highway. One-quarter mile farther is the former Colored Public School of Lynch, a two-story red brick building, constructed in 1926 and used until the late 1950s. A second area of houses and the Lynch Baptist Church are just up the road. One-quarter mile farther is the main mine entrance, portal 31, where in the early decades of the twentieth century more than one million tons of coal came through annually. There is a small parking area here and a good overview of the mine site, where much of the equipment and machinery remains. Many of the sites are within walking distance from the mine.

Directly across from the mine site is the tipple, a concrete and steel structure with a capacity of eight thousand tons of coal a day and storage facilities for more than five thousand tons. Side tracks were designed to accommodate sixty car trains, and three loading tracks ran beneath the tipple. Locomotives shuttled loaded cars from the tipple to the nearest railroad yard more than five miles away. Because of a series of fires, coal was no longer stored here after the 1940s, but it was used for loading until 1990. Coal was also taken by a conveyor system to the powerhouse, located next to the tipple, that supplied electricity to the mines and the town until 1925 when Kentucky Utilities ran power lines into the town. Nearby was the Water Filtration Plant, built in 1934, which supplied 1.5 million gallons of water a day to the town after it was enlarged in 1954.

The mine lamp house, the firehouse, the mine office, and the bathhouse are all located next to the mine and date from the early 1920s. Mine Lamp House Number 2 is being developed as a museum and has an extensive collection of mining equipment and displays. Local residents have gathered lamps, gas detecting equipment, tools, and mining memorabilia. More than two thousand electric cap lamps were issued daily from here during the 1940s. Plans are currently under way to offer complete tours of the mining area, including a trip into the mines.

The firehouse was constructed in 1920 and was the center for fire protection, ambulance service, and the Plant Protection Force of ten to fifteen men. Across the road is the Louisville & Nashville Railroad depot, one of the few brick buildings in town, which provided passenger service to Lynch from 1925 to 1956. Two passenger trains a day served the town in the 1940s. Near the depot is the district machine shop, which originally had all glass walls that were replaced with clay tiles in the 1950s.

The long building spanning the left side of the road was the bathhouse and office building. Built by Italian stonemasons, the structure is 313 feet long. The two-story section housed the main offices, and the long rambling one-story section was the bathhouse. The offices housed the general superintendent, accounting and engineering personnel, and safety, first-aid, and personnel employees. At its peak, the bathhouse could serve more than four thousand men on several shifts. It had seventy-one individual showers and more than sixty lavatories. Miners could place their street clothes on hooks, hoist them high above the room, and retrieve them at the end of their shift. In 1956, the general offices were moved next door to the present-day City Hall. This building was originally a bank and post office when built in 1917 and was the only structure in town not owned by U.S. Steel.

Just beyond City Hall, on a hillside street, is a Catholic church and the former site of the Lynch Hotel. Across the street and to the left of City Hall is the commissary, or company store. It was built in 1918 and

Homes along KY 160 at the entrance to Lynch.

Lynch is the only coal town in the region where a large historic mine site has been preserved.

remained in use until the 1950s. This was another one of U.S. Steel's showpiece buildings. The company claimed it to be the largest and most complete company store in the country. It stands 160 feet wide and more than 100 feet deep. The outside walls are made of native sandstone; the columns, floors, and beams, of concrete; and the roofing, of asbestos. This was the main store in Lynch, although there were a few small outlets and private vendors who sold staples and soft drinks. The first floor was the main storeroom for supplies, clothing, foods, meats, and drugs. Departments for women's and children's apparel were upstairs. The store help lived in the upper portion of the building, and the basement housed large ice machines and cooling facilities to store carloads of meat and produce. Like the company store in Benham, the commissary served as a social and entertainment center as well as a place for consumer goods and supplies. The commissary was open six days a week, thirteen hours a day, and employed more than one hundred clerks. It offered competitive prices and delivery services.[73] Other buildings and facilities at Lynch during its prime include theaters, churches, schools, and many more residences than exist today.

KY 160 continues up the hill and leads into upper Lynch, where many of the houses were demolished in the early 1960s.[74] At the base of Black Mountain, just beyond the Leland E. Payton football field, is a small cemetery with a few remaining tombstones that bear eastern European names. The dates indicate that many buried here died during the great influenza epidemic following World War I. Some historians have called Lynch a small kingdom, ruled by U.S. Steel, where employees and their families enjoyed a comfortable living but paid a price. The company kept a firm, orderly community and did all in its power to discourage union activities. Nonetheless, Lynch was an excellent model of welfare capitalism of the 1920s. Overall, Benham and Lynch offer the best opportunity to view intact coal towns in east Kentucky.[75]

Glossary

These select coal mining terms have been condensed from the Kentucky Coal Council's coal education glossary online. The complete glossary is available at http://www.coaleducation.org/glossary.htm.

Auger: A rotary drill that uses a screw device to penetrate, break, and then transport the drilled material (coal).

Beam: A bar or straight girder used to support a span of roof between two support props or walls.

Bed: A stratum of coal or other sedimentary material.

Bituminous coal: A middle rank coal formed by additional pressure and heat on lignite. It usually has a hi Btu value and may be referred to as "soft coal."

Coal mine: An area of land and all structures, facilities, machinery, tools, equipment, shafts, slopes, tunnels, excavations, and other property placed upon, under, or above the surface by any person, used in extracting coal from its natural deposits.

Coal reserve: Measured tonnages of coal that have been calculated to occur in a coal seam within a particular property.

Coal washing: The process of separating undesirable materials from coal based on differences in densities.

Coke: A hard, dry carbon substance produced by heating coal to a very high temperature in the absence of air.

Colliery: British name for a coal mine.

Continuous miner: A machine that constantly extracts coal while it loads it.

Conveyor: An apparatus for moving material from one

point to another in a continuous fashion. This is accomplished with an endless (looped) procession of hooks, buckets, wide rubber belts, etc.

Drift: A horizontal passage underground. A drift follows the vein, as distinguished from a crosscut that intersects it, or a level or gallery, which may do either.

Drift mine: An underground coal mine in which the entry or access is above water level and generally on the slope of a hill, driven horizontally into a coal seam.

Face: The exposed area of a coal bed from which coal is being extracted.

Fossil fuel: Any naturally occurring fuel of an organic nature, such as coal, crude oil, and natural gas.

Longwall mining: One of three major underground coal mining methods currently in use. Employs a steel plow, or rotation drum, that is pulled mechanically back and forth across a face of coal that is usually several hundred feet long. The loosened coal falls onto a conveyor for removal from the mine.

Man trip: A carrier of mine personnel by rail or rubber tire to and from the work area.

Methane: A potentially explosive gas formed naturally from the decay of vegetative matter, similar to that which forms coal. Methane, which is the principal component of natural gas, is frequently encountered in underground coal mining operations.

Pillar: An area of coal left to support the overlying strata in a mine, sometimes left permanently to support surface structures.

Preparation plant: A place where coal is cleaned, sized, and prepared for market.

Reserve: That portion of the identified coal resource that can be economically mined at the time of determination. The reserve is derived by applying a recovery factor to that component of the identified coal resource designated as the reserve base.

Roof: The stratum of rock or other material above a coal seam; the overhead surface of a coal working place. Same as "back" or "top."

Roof bolt: A long steel bolt driven into the roof of an underground excavation to support the roof, preventing and limiting the extent of roof falls.

Room and pillar method: A method of underground mining

in which approximately half of the coal is left in place to support the roof of the active mining area. Large pillars are left, while rooms of coal are extracted.

Shaft mine: An underground mine in which the main entry or access is by means of a vertical shaft.

Timber: A collective term for underground wooden supports.

Tipple: Originally the place where mine cars were tipped and emptied of their coal; although still used in that sense, the term is now more generally applied to the surface structures of a mine, including the preparation plant and loading tracks.

Tram: Used in connection with moving self-propelled mining equipment. A tramming motor may refer to an electric locomotive used for hauling loaded trips, or it may refer to the motor in a cutting machine that supplies the power for moving or tramming the machine.

Undercut: To cut below or undermine the coal face by chipping away the coal by pick or mining machine.

Waste: Rock or mineral of no value that must be removed from a mine to keep the mining scheme practical.

Working face: Any place in a mine where material is extracted during a mining cycle.

Notes

1. The Big Sandy River Valley

1. Carol Crowe-Carraco, *The Big Sandy* (Lexington: Univ. Press of Kentucky, 1979), vii, 1, 4; Mack H. Gillenwater, "Cultural and Historical Geography of Mining Settlements in the Pocahontas Coal Field of Southern West Virginia, 1880–1930" (Ph.D. diss., Univ. of Tenn., 1972), 14.
2. Harry M. Caudill, *Night Comes to the Cumberlands: A Biography of a Depressed Area* (Boston: Little, Brown & Co., 1963), 1.
3. Quoted in Edward J. Prescott, *The Story of the Virginia Coal and Iron Company, 1882–1945* (Big Stone Gap: Virginia Coal and Iron Co., 1946), 3.
4. Ibid., 437; West Virginia Writers' Program, *West Virginia: A Guide to the Mountain State* (New York: Oxford Univ. Press, 1941), 10, 12–13; Mary Jean Bowman and W. Warren Haynes, *Resources and People in East Kentucky: Problems and Potentials of a Lagging Economy* (Baltimore, Md.: Johns Hopkins Univ. Press, 1963), 27–28; Federal Writers' Project, *Kentucky: A Guide to the Bluegrass State* (New York: Harcourt, Brace and Co., 1939), 8.
5. Federal Writers' Project, *Kentucky,* 8.
6. Caudill, *Night Comes to the Cumberlands,* 63.
7. Joe Geiger, ed., *Marking Our Past: West Virginia's Historical Highway Markers* (Charleston: West Virginia Division of Culture and History, 2002), 104.
8. Ronald D Eller, *Miners, Millhands, and Mountaineers: Industrialization of the Appalachian South, 1880–1930* (Knoxville: Univ. of Tennessee Press, 1982), 86–87; Federal Writers' Project, *Kentucky,* 22; West Virginia Writers, *West Virginia,* 20.
9. Crowe-Carraco, *The Big Sandy,* 79; George F. Nielsen, ed., *1981 Keystone Coal Industry Manual* (New York, 1981), 556, 558, 632.

10. L. Martin Perry, "Coal Company Towns in Eastern Kentucky, 1854–1941," unpublished report of the Kentucky Heritage Council, Frankfort, Ky., 1991, 8; Gillenwater, "Cultural and Historical Geography," 17.

11. Federal Writers' Project, *Kentucky,* 16.

12. Ibid.; West Virginia Writers' Program, *West Virginia,* 19.

13. Eller, *Miners, Millhands, and Mountaineers,* 17; William David Deskins, *Pike County: A Very Different Place* (Pikeville, Ky.: Printing by George, 1994), 157.

14. Eller, *Miners, Millhands, and Mountaineers,* 22, 23.

15. Altina L. Waller, *Feud: Hatfields, McCoys, and Social Change in Appalachia, 1860–1900* (Chapel Hill: Univ. of North Carolina Press, 1988), 212–13.

16. Eller, *Miners, Millhands, and Mountaineers,* 8.

17. An excellent study of settlement names is Robert M. Rennick, *Kentucky Place Names* (Lexington: Univ. Press of Kentucky, 1984).

18. Eller, *Miners, Millhands, and Mountaineers,* 67, 78.

19. Thomas W. Dixon Jr., *Chesapeake and Ohio in the Coalfields* (Clifton Forge, Va.: Chesapeake & Ohio Historical Society, 1995), 1–3.

20. Ed King, *Norfolk & Western in the Appalachians: From the Blue Ridge to the Big Sandy* (Waukesha, Wis.: Kalmbach Books, 1998), 5–7.

21. Ibid., 95.

22. Ibid.; Thomas H. Garvey, *The Last Steam Railroad in America* (New York: Harry N. Abrams, 1995), 6–9.

23. King, *Norfolk & Western in the Appalachians,* 11.

24. Ibid., 80.

25. Hugh Wolfe and Ed Wolfe, *Appalachian Coal Hauler: The Interstate Railroad's Mine Runs and Coal Trains* (Lynchburg, Va.: TLC Publishing, 2001), 6–7.

26. Eller, *Miners, Millhands, and Mountaineers,* 150–51.

27. Thomas W. Dixon, *Appalachian Coal Mines and Railroads* (Lynchburg, Va.: TLC Publishing, 1994), 6.

28. *Pike County News,* 5 Dec. 1924, 16 Aug. 1928, 4 Apr. 1929, 9, 16 July 1931, 25 June 1936.

29. Ibid., 26 Aug. 1924.

30. See for example, *Kentucky Atlas and Gazetteer* (Yarmouth, Maine: DeLorme, 1997).

31. Perry, "Coal Company Towns," 7–10; Dept. of the Interior, National Parks Service, *Coal Fields, Communities and Change* (Philadelphia: National Parks Service Mid-Atlantic Regional Office, 1991); *A Coal Heritage Study: A Study of Coal Mining and Related Resources in Southern West Virginia* (Philadelphia: National Parks Service, Mid-Atlantic Regional Office, 1993); Sharon A. Brown, *A Historic Resource Study: Kaymoor, New River Gorge National Park, West Virginia* (Washington, D.C.: U.S. Dept. of the Interior, 1990).

2. A General History of the Big Sandy River Valley

1. Henry P. Scalf, *Kentucky's Last Frontier,* 2d ed. (Pikeville, Ky.: Pikeville College Press, 1972), 435; William Webb, *The C & O Mounds* (Lexington: Univ. Press of Kentucky, 1942), 320; Otis K. Rice, *West Virginia: A History* (Lexington: Univ. Press of Kentucky, 1985), 12–13.
2. Crowe-Carraco, *The Big Sandy,* 10. The Totero were probably related to the Siouian-speaking Tutelo of southwestern Virginia. The latter group later migrated to North Carolina.
3. Rice, *West Virginia,* 12–13.
4. Scalf, *Kentucky's Last Frontier,* 35, 36.
5. Ibid., 37. Both Scalf and Crowe-Carraco report that Gist traveled along Elkhorn Creek, a tributary of the Big Sandy River. See Crowe-Carraco, *The Big Sandy,* 11.
6. Crowe-Carraco, *The Big Sandy,* 17–18; Scalf, *Kentucky's Last Frontier,* 70–76; C. Mitchell Hall, *Johnson County, Kentucky: A History of the County and Genealogy of Its People up to the Year 1927* (Louisville, Ky.: Standard Press, 1928), 1:55–56.
7. Thomas D. Clark, *A History of Kentucky* (Ashland, Ky.: Jesse Stuart Foundation, 1992), 63.
8. Crowe-Carraco, *The Big Sandy,* 20; Scalf, *Kentucky's Last Frontier,* 164–67.
9. Crowe-Carraco, *The Big Sandy,* 19.
10. Otis K. Rice, *The Allegheny Frontier: West Virginia Beginnings, 1730–1830* (Lexington: Univ. Press of Kentucky, 1970), 67–68; Clark, *History of Kentucky,* 76–77.
11. Waller, *Feud,* 22.
12. David Hackett Fischer, *Albion's Seed* (New York: Oxford Univ. Press, 1991), 325.
13. Crowe-Carraco, *The Big Sandy,* 51.
14. John David Preston, *The Civil War in the Big Sandy Valley of Kentucky* (Baltimore Md.: Gateway Press, 1984), 18–19.
15. Crowe-Carraco, *The Big Sandy,* 48.
16. Ibid., 49; Otis Rice, *The Hatfields and the McCoys* (Lexington: Univ. Press of Kentucky Press, 1982), 10–11; Waller, *Feud,* 32–33.
17. Crowe-Carraco, *The Big Sandy,* 64.
18. Eller, *Miners, Millhands, and Mountaineers,* 87–91.
19. Crowe-Carraco, *The Big Sandy,* 68–69.
20. George Wolfford, *Lawrence County: A Pictorial History* (Ashland, Ky.: W.W.W. Co., 1972), 35; William Ely, *The Big Sandy Valley: A History of the People and Country from the Earliest Settlement to the Present Time* (Cattlettsburg, Ky.: Genealogical Publishing Co., 1969), 308–9.

21. Scalf, *Kentucky's Last Frontier*, 327.
22. Eller, *Miners, Millhands, and Mountaineers*, 128.
23. Ibid., 58–59.
24. For more on John C.C. Mayo, see Carolyn Clay Turner and Carolyn Hay Traum, *John C.C. Mayo: Cumberland Capitalist* (Pikeville, Ky.: Pikeville College Press, 1983).
25. Caudill, *Night Comes to the Cumberlands*, 74.
26. Dept. of the Interior, *A Coal Heritage Study*, 52–53.
27. Caudill, *Night Comes to the Cumberlands*, 140–45.
28. See Joe William Trotter Jr., *Coal, Class, and Color: Blacks in Southern West Virginia, 1915–1932* (Urbana: Univ. of Illinois Press, 1990).
29. Eller, *Miners, Millhands, and Mountaineers*, 193.
30. Crandall A. Shifflett, *Coal Towns: Life, Work, and Culture in Company Towns of Southern Appalachia, 1880–1930* (Knoxville: Univ. of Tennessee Press, 1991), 117.
31. David A. Corbin, *Life, Work, and Rebellion in the Coalfields: The Southern West Virginia Miners, 1880–1922* (Urbana: Univ. of Illinois Press, 1981), 185, 181.
32. Daniel P. Jordan, "The Mingo War: Labor Violence in the Southern West Virginia Coal Fields, 1919–1922," in *Essays in Southern Labor History: Selected Papers, Southern Labor History Conference, 1976*, ed. Gary M. Fink and Merl E. Reed (Westport, Conn.: Greenwood Press, 1977), 106–7.
33. Lon K. Savage, *Thunder in the Mountains: The West Virginia Mine War, 1920–1921* (Pittsburgh: Univ. of Pittsburgh Press, 1990), 16–17.
34. Ibid., 19–24.
35. Ibid., 14–16; Corbin, *Life, Work, and Rebellion*, 201.
36. Howard B. Lee, *Bloodletting in Appalachia: The Story of West Virginia's Four Major Mine Wars and Other Thrilling Incidents of Its Coal Fields* (Morgantown: West Virginia Univ. Press, 1969), 53–56; Savage, *Thunder in the Mountains*, 55.
37. John W. Bond, "National Historic District Landmark Nomination Form, Matewan Historic District," sec. 1, 19; West Virginia State Archives, "West Virginia's Mine Wars," at http://www.wvculture.org/history/minewars.html.
38. Savage, *Thunder in the Mountains*, 57–59.
39. Ibid., 69–70.
40. Lee, *Bloodletting in Appalachia*, 91–92.
41. Clayton D. Laurie, "The United States Army and the Return to Normalcy in Labor Dispute Intervention: The Case of the West Virginia Coal Mine Wars, 1920–1921," at http:/www.wvculture.org/history/journal_wvh/wvh50-1.html.
42. Summarized from Savage, *Thunder in the Mountains*, 111–60.
43. Ibid., 166–67.

44. Shifflett, *Coal Towns*, 117; Crowe-Carraco, *Big Sandy*, 104.
45. Keith Dix, *What's a Coal Miner to Do? The Mechanization of Coal Mining* (Pittsburgh: Univ. of Pittsburgh Press, 1988), 169, 175.
46. Thomas D. Matijasic, "The Ku Klux Klan in the Big Sandy Valley of Kentucky," *Journal of Kentucky Studies* 3 (Sept. 1993): 7; Brown, *Historic Resource Study*, 26.
47. Rice, *West Virginia*, 234.
48. Dix, *What's a Coal Miner to Do?* 191–92.
49. Caudill, *Night Comes to the Cumberlands*, 264–65.
50. Ibid., 258; Harry M. Caudill, *Theirs Be the Power: The Moguls of Eastern Kentucky* (Urbana: Univ. of Illinois Press, 1983), 128; Crowe-Carraco, *The Big Sandy*, 111–12.
51. *Harlan (Ky.) Heritage*, 28 Feb. 1984.
52. Ibid.
53. For more on Gov. Bert T. Combs, see George W. Robinson, "Bert T. Combs," in *Kentucky's Governors, 1792–1985*, ed. Lowell H. Harrison (Lexington: Univ. Press of Kentucky, 1985), 166–70; George W. Robinson, ed., *Bert Combs, the Politician: An Oral History* (Lexington: Univ. Press of Kentucky, 1991).
54. Rice, *West Virginia*, 286.
55. For more on the War on Poverty, see John M. Glen, "The War of Poverty in Appalachia—A Preliminary Report," *Register of the Kentucky Historical Society* 20 (winter 1989): 40–57.
56. Crowe-Carraco, *The Big Sandy*, 116.
57. *Lexington (Ky.) Herald-Leader*, 1 July 1996.
58. Crowe-Carraco, *The Big Sandy*, 121.
59. Kristin Layne Szakos, "Kentuckians for the Commonwealth," in *The Kentucky Encyclopedia*, ed. John E. Kleber (Lexington: Univ. Press of Kentucky, 1992), 489; Rice, *West Virginia*, 285.
60. Penny Loeb, "Shear Madness," *U.S. News and World Report* 123 (11 Aug. 1977): 28–30.

3. Coal, Coal Mining, Coal Towns, and Structures

1. Priscilla Long, *Where the Sun Never Shines: A History of America's Bloody Coal Industry* (New York: Paragon House, 1989), 4–5; Phil Conley, *History of the West Virginia Coal Industry* (Charleston, W.Va.: Education Foundation, 1960), 1–3.
2. Nielsen, *1981 Keystone Coal Manual*, 548; Shifflett, *Coal Towns*, 30.
3. Shifflett, *Coal Towns*, 30, 93; Michael E. Workman, Paul Salstrom, and Philip W. Ross, *Northern West Virginia Coal Fields: Historical Context* (Morgantown, W.Va.: West Virginia Univ. Press, 1994), 20.

4. Production figures provided by the U.S. Dept. of Energy and the West Virginia Office of Miners Health, Safety, and Training.

5. Long, *Where the Sun Never Shines,* 7–10.

6. John R. Stilgoe, *Common Landscape of America, 1580 to 1845* (New Haven, Conn.: Yale Univ. Press, 1982), 299.

7. Stilgoe, *Common Landscape of America,* 299; Long, *Where the Sun Never Shines,* 20–22.

8. John R. Stilgoe, *Metropolitan Corridor: Railroads and the American Scene* (New Haven, Conn.: Yale Univ. Press, 1983), 117–19; Shifflett, *Coal Towns,* 28.

9. Shifflett, *Coal Towns,* 81, 87; Long, *Where the Sun Never Shines,* 23.

10. Dix, *What's a Coal Miner to Do?* 2–3; Shifflett, *Coal Towns,* 81–105.

11. Dix, *What's a Coal Miner to Do?* 3.

12. David Zegeer, "Mechanization of the Coal Industry in Appalachia," unpublished paper, 1999, 8; Long, *Where the Sun Never Shines,* 82.

13. Shifflett, *Coal Towns,* 89.

14. Dix, *What's a Coal Miner to Do?* 5–18.

15. Workman, Salstrom, and Ross, *Northern West Virginia Coalfields,* 24–25; Shifflett, *Coal Towns,* 105–7.

16. Shifflett, *Coal Towns,* 106–8.

17. Zegeer, "Mechanization of the Coal Industry," 8–9.

18. Dix, *What's a Coal Miner to Do?* 37; Zegeer, "Mechanization of the Coal Industry," 4–5.

19. Dix, *What's a Coal Miner to Do?* 77–79.

20. Caudill, *Night Comes to the Cumberlands,* 261–62.

21. Corbin, *Life, Work, and Rebellion,* 13; Crowe-Carraco, *The Big Sandy,* 87.

22. Dix, *What's a Coal Miner to Do?* 88; Rice, *West Virginia,* 237; Kleber, *Kentucky Encyclopedia,* 640.

23. Dix, *What's a Coal Miner to Do?* 88.

24. Kleber, *Kentucky Encyclopedia,* 640.

25. Rice, *West Virginia,* 237; Kleber, *Kentucky Encyclopedia,* 640.

26. Perry, "Coal Company Towns," 37.

27. Robert F. Munn, "The Development of Model Towns in the Bituminous Coal Fields," *West Virginia History* 9 (spring 1979): 243–44; Margaret Ripley Wolfe, "Putting Them in Their Place: Industrial Housing in Southern Appalachia, 1900–1930," *Appalachian Heritage* 7 (summer 1979): 28; Perry, "Coal Company Towns," 37.

28. Eller, *Miners, Millhands, and Mountaineers,* 183, 185–86, 192.

29. Gillenwater, "Cultural and Historical Geography," 43–44.

30. Richard V. Francavaglia, *Hard Places: Reading the Landscape of America's Historic Mining Districts* (Iowa City: Univ. of Iowa Press, 1991), 5, 33.

31. Francavaglia, *Hard Places,* 31; A. F. Huebner, "Houses for Mine Villages," *Coal Age* 12 (27 Oct. 1917): 717–18; George H. Miller, "Plan Your Town as

Carefully as You Would Your Plant," *Coal Age* 14 (20 July 1918): 130; U.S. Coal Commission, *Report of the United States Coal Commission,* 5 vols. (Washington, D.C., 1923), 3:1428–29; Corbin, *Life, Work, and Rebellion,* 67–68.

32. Eller, *Miners, Millhands, and Mountaineers,* 185; Corbin, *Life, Work, and Rebellion,* 68.

33. Gillenwater, "Cultural and Historical Geography," 57. Gillenwater identified four basic designs found in the Pocahontas coal field. A linear design, with short streets built off of the main road appears to be the most common in the Big Sandy River Valley.

34. Caudill, *Night Comes to the Cumberlands,* 99.

35. Margaret Crawford, *Building the Workingman's Paradise: The Design of American Company Towns* (New York: Verso, 1995), 61–78.

36. Eller, *Miners, Millhands, and Mountaineers,* 190; Crawford, *Workingman's Paradise,* 160–62: Corbin, *Life, Work, and Rebellion,* 122–23.

37. Crawford, *Workingman's Paradise,* 45; Wolfe, "Putting Them in Their Places," 27; Munn, "Development of Model Towns," 247.

38. Shifflett, *Coal Towns,* 37.

39. Munn, "Development of Model Towns," 248; Wolfe, "Putting Them in Their Places," 29–30. U.S. Oil became Island Creek Coal Company in 1915.

40. Wolfe, "Putting Them in Their Place," 29.

41. Eller, *Miners, Millhands, and Mountaineers,* 14.

42. R. H. Hamill, "Design of Buildings in Mining Towns," *Coal Age* 11 (14 June 1917): 1045; Gillenwater, "Cultural and Historical Geography of Mining Settlements," 70.

43. Corbin, *Life, Work, and Rebellion,* 67; Eller, *Miners, Millhands, and Mountaineers,* 184.

44. *Coal Age* 12 (27 Oct. 1917): 717.

45. Huebner, "Houses for Mine Villages," 717–18; Hamill, "Design of Buildings," 1045, 1048; Brown, *Historic Resource Study,* 66.

46. Robert Schweitzer and Michael W. R. Davis, *America's Favorite Homes: Mail Order Catalogues as a Guide to Popular Early Twentieth Century Houses* (Detroit: Wayne State Univ. Press, 1990), 109–12; Perry, "Coal Company Towns," 32.

47. Huebner, "Houses for Mine Villages," 719.

48. U.S. Coal Commission, *Report of the U.S. Coal Commission,* 3:1429–32; *Newsweek* 29 (May 5, 1947): 56.

49. Jakle, Bastian, and Meyer, *Common Houses,* 171; Huebner, "Houses for Mine Villages," 718–19; Francaviglia, *Hard Places,* 102; Corbin, *Life, Work, and Rebellion,* 124.

50. Brown, *Historic Resource Study,* 67–68; Trotter, *Coal, Class, and Color,* 129; Corbin, *Life, Work, and Rebellion,* 67.

51. Trotter, *Coal, Class, and Color,* 128–29.

52. Shifflett, *Coal Towns,* 191.
53. Gillenwater, "Cultural and Historical Geography," 96–97.
54. Brown, *Historic Resource Study,* 79.
55. Gillenwater, "Cultural and Historical Geography," 99.
56. Shifflett, *Coal Towns,* 176–89; Brown, *Historic Resource Study,* 74; Gordon Dodrill, comp., *Twenty Thousand Coal Company Stores in the United States, Mexico, and Canada* (Pittsburgh: Univ. of Pittsburgh Press, 1971).
57. Price Fishback, "Did Miners Owe Their Souls to the Company Store?" *Journal of Economic History* 46 (Dec. 1986): 1011–29.
58. Shifflett, *Coal Towns,* 177–78.
59. Gillenwater, "Cultural and Historical Geography," 91.
60. Caudill, *Night Comes to the Cumberlands,* 145.
61. Francaviglia, *Hard Places,* 50.
62. Dixon, *Appalachian Coal Mines,* 4–5.
63. Shifflett, *Coal Towns,* 93–94; Paul W. Thrush, *A Dictionary of Mining, Mineral, and Related Terms* (Washington, D.C.: U.S. Bureau of Mines, 1968), 232–33.
64. Francaviglia, *Hard Places,* 55; Brown, *Historic Resource Study,* 18.
65. Gillenwater, "Cultural and Historical Geography," 92.
66. Brown, *Historic Resource Study,* 73.
67. Gillenwater, "Cultural and Historical Geography," 101.
68. Monica B. Hawley, "Boldman Bridge (KY 1384 Suspension Bridge)," *Historic American Engineering Record,* 1984.
69. Stilgoe, *Metropolitan Corridor,* 197–98.
70. Ibid., 198.
71. Perry, "Coal Company Towns," 9.

4. Up the Tug Fork through the Coal Towns of West Virginia

1. All population estimates are from the Rand-McNally *2002 Commercial Atlas & Marketing Guide* (Chicago: Rand-McNally 2002).
2. West Virginia Writers' Program, *West Virginia,* 479; Dept. of Interior, *Coal Heritage Study,* 5–6, 11.
3. Shifflett, *Coal Towns,* 41.
4. Ibid., 44–45.
5. Savage, *Thunder in the Mountains,* 30, 39; Corbin, *Life, Work, and Rebellion,* 204.
6. West Virginia Writers', *West Virginia,* 478.
7. Mae Stallard, "A Walking Tour of Williamson," Tug Valley Chamber of Commerce, n.d.

8. E. Lee North, *The 55 West Virginias: A Guide to the State's Counties* (Morgantown, W.Va.: West Virginia Univ. Press, 1998), 48.

9. Conley, *History of West Virginia,* 249–50; Walter R. Thurmond, *The Logan Coal Field of West Virginia: A Brief History* (Morgantown: West Virginia Univ. Library, 1964), 58.

10. Conley, *History of West Virginia,* 250.

11. "Miners' Town," *Goldenseal* 8 (winter 1982): 55–57.

12. Shifflett, *Coal Towns,* 234 n.

13. *Coal Age* 39 (Apr. 1934): 121–22.

14. Carl Agsten, "National Register of Historic Places Registration Form, Chafin House," sec. 7, 1, 3.

15. See http://mymountain.com/path1/park/chief.htm.

16. Penny Loeb, "Coal Activists Stir up Dust in West Virginia," *U.S. News and World Report* 123 (13 Oct. 1997): 8; idem, "Shear Madness," *U.S. News and World Report* 123 (11 Aug. 1997): 27–36.

17. Michael Gioulis, Paul McAllister, and Stacy Sone, "National Register of Historic Places Registration Form, Matewan Historic District," sec. 8, 7.

18. Gioulis, McAllister, and Sone, "Matewan Historic District," sec. 7, 1.

19. Dodrill, *Twenty Thousand Company Stores,* 213.

20. Richard D. Lunt, *Law and Order vs. the Miners: West Virginia, 1907–1933* (Hamden, Conn.: Archon Books, 1979), 93–94.

21. Ibid., 92.

22. Savage, *Thunder in the Mountains,* 37.

23. Lunt, *Law and Order,* 95.

24. Author correspondence with Rev. Curtis Stallar, Red Jacket, W.Va.

25. Gillenwater, "Cultural and Historical Geography," 71, 77.

26. Information provided by Mr. David Reynolds, Matewan Development Center, Matewan, W.Va.

27. McGehee and McGuire, *A Century of Stewardship,* 75.

28. Stacy Sone, "National Register of Historic Places Multiple Property Documentation Form, Coal Company Stores in McDowell County," sec. E, 2.

29. Kim A. Valentine, "National Register of Historic Places Registration Form, Welch Commercial Historic District," sec. 7, 1; sec. 8, 4.

30. West Virginia Writers' Program, *West Virginia,* 476.

31. Geiger, ed., *Marking Our Past,* 96.

32. Charles E. Turley, "National Register of Historic Places register Form, McDowell County Courthouse," sec. 8, 2, 4.

33. Stuart McGehee, "Gary: A First Class Mining Operation," unpublished paper, Eastern Regional Coal Archives, Craft Memorial Library, Bluefield, W.Va.

34. McGehee, "Gary," 4.

35. Peter Westleigh, "Safety the First Consideration," *Coal Age* 3 (1 Mar. 1913): 349.

36. Ibid.

37. McGehee, "Gary," 6.

38. Ibid., 5–6; *Welch (W.Va.) Daily News,* 3 June 1983.

39. *Bluefield (W.Va.) Daily Telegraph,* 24 June 1991.

40. *Washington Post,* 17 Dec. 1986.

41. *Welch (W.Va.) Daily News,* 3 June 1938; Stuart McGehee, "Historic Coalwood," *Goldenseal* 27 (summer 2001): 51, 52.

42. "The Consolidation Coal Company: An Industrial Giant," *West Virginia Review* 20 (Oct. 1931): 74–76.

43. Ibid.; *Bluefield (W.Va.) Daily Telegraph,* 31 Dec. 1989.

44. *Bluefield (W.Va.) Daily Telegraph,* 31 Dec. 1989.

45. Ibid., 19 Dec. 1991.

46. Stacy Sone, "National Register of Historic Places Registration Form, Carter Coal Company Store," sec. 7, 1.

47. Ibid., sec. 7, 2–3; Dept. of Interior, *Coal Heritage Study,* 70.

48. *Welch (W.Va.) Daily News,* 3 June 1958.

49. Ibid.

50. *Bluefield (W.Va.) Daily Telegraph,* 12 Dec. 1990.

51. Kathy Matney, "Too Many Tragedies: Survivor Account Details of the 1940 Bartley Mine Disaster," *Goldenseal* 4 (Oct–Dec. 1978): 50–56; Stuart McGehee, "Pioneering Coal People," *Coal People Magazine* 15 (May 1990): 11–13; *Welch (W.Va.) Daily News,* 3 June 1958.

52. West Virginia Writers' Program, *West Virginia,* 474–75.

53. Stuart McGehee, comp., "A Busy Time in McDowell History: Looking Back with John J. Lincoln," *Goldenseal* 15 (fall 1989): 56–64; "73 Years Ago Today," *Welch (W.Va.) Daily Telegraph,* 19 Aug. 1988.

54. McGehee, comp., "A Busy Time in McDowell History," 58, 56.

55. Joan D. Semonco, "Elkhorn West Virginia 24831," unpublished paper, Eastern Regional Coal Archives, Craft Memorial Library, Bluefield, W.Va.

56. Semonco, "Elkhorn West Virginia 24831," 3.

57. Stacy Sone, "John J. Lincoln House," National Register of Historic Places Registration Form, sec. 7, 1–2.

58. Conley, *History of West Virginia,* 227–30.

59. Beth Hager, "National Register of Historic Places Inventory—Nomination Form, Bramwell Historic District," sec. 7, 1, 5.

60. Michael Gioulis, "National Register of Historic Places Registration, Bramwell Additions Historic District," sec. 7, 7–8.

61. Trotter, *Coal, Class, and Color,* 45.

62. Gioulis, "Bramwell Additions Historic District," sec. 8, 12–13.

63. Geiger, ed., *Marking Our Past,* 100.

64. Ibid., 101.

65. Gillenwater, "Cultural and Historical Geography," 115, 117, 118.

66. Virginia Historic Landmark Commission Survey, "National Register of Historic Places Inventory—Nomination Form, Pocahontas Historic District," sec. 7, 2.

67. Jones, *Early Mining in Pocahontas, Virginia,* 66.

68. Miscellaneous information graciously provided by Greg Jones, Treasurer, Pocahontas, Va.

69. Virginia Historic Commission, "Pocahontas Mine No. 1," National Historic Landmark Nomination, sec. 1, 5.

70. West Virginia Writers' Program, *West Virginia,* 472.

71. Stuart McGehee, "The History of Bluefield, West Virginia," at http://www.ci.bluefield.wv.us/history/stewart.html; McGehee and McGuire, *Century of Stewardship,* 48.

72. Bluefield Visitors and Convention Bureau, "West Virginia's Historic Downtown Bluefield Walking Tour," n.d.

5. Coal Towns in East Kentucky

1. Kleber, *Kentucky Encyclopedia,* 456.

2. Donald E. Rist, *Kentucky Iron Furnaces of the Hanging Rock Iron Region* (Ashland, Ky.: Hanging Rock Press, 1974), 72–74; Rennick, *Place Names,* 205. The post office at Mt. Savage closed in 1916.

3. Dianne Wells, comp., *Roadside History: A Guide to Kentucky Highway Markers* (Frankfort: Kentucky Historical Society, 2002), 74.

4. Rennick, *Kentucky Place Names,* 258; Perry, "Mining Towns," 41.

5. Kleber, *Kentucky Encyclopedia,* 108–9.

6. *Lexington (Ky.) Herald-Leader,* 28 Dec. 1998.

7. Ibid., 22 May 1999, 2 Sept. 2000; Dixon, *Chesapeake & Ohio in the Coalfields,* 5–6, 50; Eugene L. Huddleston, "Russell Terminal in the Days of Steam," *Chesapeake & Ohio Historical Magazine* 33 (Oct. 2001): 6.

8. Kleber, *Kentucky Encyclopedia,* 503, 620.

9. Wells, *Roadside History,* 88, 90.

10. Ibid., 114, 110.

11. Turner and Traum, *John C.C. Mayo,* 83.

12. Ashland Dept. of Recreation and Parks, "Ashland Historic Tour," brochure, 1979.

13. Carolyn Bills, "Jesse Stuart 1906–1984," at: http://www.ncteamricancollection.org/litmap/stuart_jesse_ky.htm.

14. Wells, *Roadside History,* 71; Kleber, *Kentucky Encyclopedia,* 538.

15. Ely, *The Big Sandy Valley*, 308–12; Wolfford, *Lawrence County*, 13, 35; Regina Tackett, Patricia Jackson, and Janice Thompson, eds., *History of Lawrence County, Kentucky* (Dallas: Curtis Media Corp., 1991), 224.

16. Ely notes that this railroad began as a narrow gauge but was transformed to standard gauge railroad before its completion in 1882. Also see Clayton R. Cox, comp., *Appalachian Crossroads: Descendants of Hezekiah Sellards* (Baltimore, Md.: Gateway Press, 1977), 513–14.

17. Sharon H. Kinner and Carolyn H. Hale, "Peach Orchard Coal Company," in *History of Lawrence County, Kentucky*, ed. Regina Tackett, Patricia Jackson, and Janice Thompson (Dallas: Harlan Press, 1991), 64–65.

18. Wolfford, *Lawrence County*, 38.

19. Ibid., 39.

20. Kinner and Hale, "Peach Orchard Coal Company," in Tackett, Jackson, and Thompson, *History of Lawrence County*, 65.

21. Cox, *Appalachian Crossroads*, 513; Tackett, Jackson, and Thompson, *History of Lawrence County*, 224.

22. Kleber, *Kentucky Encyclopedia*, 613.

23. Doug Cantrell, "Himlerville: Hungarian Cooperative Mining in Kentucky," *Filson Club History Quarterly* 66 (Oct. 1992): 517–18; J. R. Haworth, "Hungarians Successfully Conduct Co-Operative Mine in Kentucky, Having Two-Million Dollars Invested," *Coal Age* 20 (15 Sept. 1921): 412.

24. *Frankfort (Ky.) State Journal*, 23 Mar. 1980.

25. *Coal Age* 20 (15 Sept. 1921): 413.

26. *Frankfort (Ky.) State Journal*, 23 Mar. 1980.

27. Additional information about Himlerville can be found in Crowe-Carraco, *The Big Sandy*, 93–94; *Coal Age* 20 (15 Sept. 1921): 413; Eller, *Miners, Millhands, and Mountaineers*, 191–92. Eller mistakenly refers to Himler as Henrich Himler; Rennick, *Kentucky Place Names*, 16.

28. *Frankfort (Ky.) State Journal*, 23 Mar. 1980.

29. Kleber, *Kentucky Encyclopedia*, 476.

30. Federal Writers' Project, *Kentucky*, 239.

31. Scalf, *Kentucky's Last Frontier*, 123–24; Hall, *Johnson County*, 1:174.

32. Turner and Traum, *John C.C. Mayo*, 4–5.

33. *Paintsville (Ky.) Herald*, 4 Sept. 1924.

34. Wells, *Roadside History*, 190.

35. *1989 Mayo State Vocational-Technical School Catalog*, Paintsville, Ky., 1989, 3.

36. Johnson County Historical and Genealogical Society, *Historic Sites of Paintsville and Johnson County, Kentucky* (Frankfort, Ky.: Kentucky Heritage Council, 1985), 26.

37. Turner and Traum, *John C.C. Mayo*, 9–10.

38. Dana Jean Rucker, "Van Lear: The Rise and Decline of a Coal Mining Town," in *The Research Historians' Essays on Kentucky History*, vol. 13,

unpublished, Johnson County Public Library, Paintsville, Ky.; James Vaughan, *Blue Moon over Kentucky: A Biography of Kentucky's Troubled Highlands* (n.p., 1985), 45; Turner and Traum, *John C.C. Mayo*, 45.

39. Ibid., 96; Charles E. Beachley, *History of Consolidation Coal Company, 1864–1934* (New York: Consolidation Coal Co., 1934), 54–55. Consolidation Coal's No. 151, 152, 153, 154, and 155 mines were located at Van Lear.

40. *Bankmule* 1 (Mar. 1984): 1–2.

41. C. Mitchell Hall, *Johnson County Kentucky: A History of the County and Genealogy of Its People up to the Year 1927,* 2 vols. (Louisville Ky.: Standard Press, 1928), 1:262.

42. Thomas D. Matijasic, "The Ku Klux Klan in the Big Sandy Valley of Kentucky," *Journal of Kentucky Studies* 10 (Sept. 1993): 77.

43. *Bankmule* 6 (Sept. 1984): 8.

44. Jeanette Knowles, "Henry Skaggs," *Bankmule* 1 (July 1984): 3.

45. Joyce Meade, "Frank Campigotti," *Bankmule* 2 (July 1985): 17.

46. Charlene Conley, "A Tale about 'Bankmules' or What Is Was a Football Game!" *Bankmule* 2 (Dec. 1985): 4.

47. *Coal Age* 40 (Aug. 1935): 102.

48. Vaughn, *Blue Moon,* 96; Rucker, "Van Lear," 243.

49. Johnson County Historical Society, *Historic Sites of Paintsville,* 31.

50. *Bankmule* 1 (July 1984): 10–11.

51. Father Frank C. Osburg, "Saint Casimir Mission Center—Where It All Began," *Bankmule* 3 (Dec. 1986): 5.

52. Scalf, *Kentucky's Last Frontier,* 329.

53. Ibid., 329–30.

54. Ibid., 331.

55. C. Mitchell Hall, *Jenny Wiley Country: A History of the Big Sandy Valley in Kentucky's Eastern Highlands and Genealogy of the Region's People* (Kingsport, Tenn.: Professional Press, 1972–79), 1:349–50.

56. Scalf, *Kentucky's Last Frontier,* 104–6; Henry P. Scalf, *Floyd County Sesquicentennial, 1800–1950* (Prestonsburg, Ky.: Floyd Press, 1950), 25.

57. The ledger of Floyd County merchant John Graham gives a good indication of the importance of the fur trade in the Big Sandy Valley during the early nineteenth century. See Nancy O'Malley, "The DeRossett-Johns Site Archaeological Report," no. 243, Kentucky Heritage Council, Frankfort, Ky., 1990, 9.

58. Scalf, *Kentucky's Last Frontier,* 197–98.

59. Ibid., 200–201. The Peach Orchard mines to the west of Prestonsburg should not be confused with the Peach Orchard mining operations in Lawrence County.

60. Crowe-Carraco, *The Big Sandy,* 64.

61. Ibid., 67.

62. Paulin Archer Burchett, "A Brief History of Prestonsburg 1800–1930," in Traum, *History of Floyd County,* 67.

63. Delmas Saunders, "Colonial Hollow," in Traum, *History of Floyd County*, 35.

64. Traum, *History of Floyd County*, 67.

65. William Jennings Martin and Alice O. Martin, "Garrett," in Traum, *History of Floyd County*, 44; Rennick, *Kentucky Place Names*, 114.

66. Rudolph Spencer interview, 8 Sept. 1982, Appalachia Oral History Project, Alice Lloyd College, Pippa Passes, Ky.

67. Crowe-Carraco, *The Big Sandy*, 104.

68. Dodrill, *Twenty Thousand Company Stores*, 73.

69. Phyllis B. Honshell, "Wayland," in Traum, *History of Floyd County*, 72.

70. Traum, *History of Floyd County*, 77; Dept. of the Census, *Nineteenth Census of the United States*, 1:11.

71. Kleber, *Kentucky Encyclopedia*, 213.

72. Traum, *History of Floyd County*, 38; Rennick, *Kentucky Place Names*, 85. Rennick notes that the origins of the town's name are not known. Another story notes a drift mine in the area.

73. Dodrill, *Twenty Thousand Company Stores*, 83.

74. Thomas Matijasic interview with Raymond Wright, Dec. 1995, McDowell, Ky.

75. Traum, *History of Floyd County*, 39.

76. Rennick, *Kentucky Place Names*, 183.

77. Traum, *History of Floyd County*, 62.

78. *Big Sandy News*, 8 Sept. 1916; Eller, *Miners, Millhands, and Mountaineers*, 144.

79. E. R. Price, "Wheelwright Program Brings Increased Safety and Efficiency with Lower Operating Costs," *Coal Age* 37 (10 Dec. 1932): 425; Lewis M. Williams, *The Transformation of a Coal Mining Town* (Chicago: Inland Steel Co., n.d.), 3; Traum, *History of Floyd County*, 80; Shifflett, *Coal Towns*, 59, 187.

80. *Louisville (Ky.) Courier-Journal*, Dec. 21, 1992.

81. *Coal Age* 37 (10 Dec. 1932): 426; Williams, *Transformation of a Coal Mining Town*, 10.

82. *Louisville (Ky.) Courier-Journal*, 21 Dec. 1992.

83. "Wheelwright, Kentucky Is Ultra-Modern," in *Kentucky* 12 (winter 1949): 15; Williams, *Transformation of a Coal Town*, 6–7; Shifflett, *Coal Towns*, 56; *Pike County News*, 10 June 1937, 28 May 1942, 8 Oct. 1951.

84. Shifflett, *Coal Towns*, 57, 60; Williams, *Transformation of a Coal Town*, 19, 23.

85. Shifflett, *Coal Towns*, 60, 64.

86. *Louisville (Ky.) Courier-Journal*, 21 Dec. 1992 (from a 1987 interview).

87. Shifflett, *Coal Towns*, 23.

88. *Louisville (Ky.) Courier-Journal*, 21 Dec. 1992.

89. U.S. Dept. of Interior, Coal Mines Administration, *A Medical Survey of the Bituminous Coal Industry*, Washington, D.C., 1947.

90. Dept. of the Census, *Eighteenth Decennial Census of the United States,* 1:17.

91. Shifflett, *Coal Towns,* 59.

92. Dixon, *Chesapeake & Ohio in the Coalfields,* 97.

93. *Lexington (Ky.) Herald-Leader,* 15 Dec. 1996.

94. Ibid.

95. Kleber, *Kentucky Encyclopedia,* 722.

96. Crowe-Carraco, *The Big Sandy,* 24; Deskins, *Pike County,* 9.

97. Federal Writers' Project, *Kentucky,* 241.

98. Crowe-Carraco, *Big Sandy,* 73, 85; *Pike County News,* 4 Jan. 1936, 16 June 1938, 20 June 1940.

99. Deskins, *Pike County,* 171.

100. Crowe-Carraco, *Big Sandy,* 115.

101. Rice, *The Hatfields and McCoys,* 4–5.

102. Waller, *Feud,* 40–43. For a short summary of Waller's findings, see Altina L. Waller, "Feuding and Modernization in Appalachia: The Hatfields and McCoys," *Register of the Kentucky Historical Society* 20 (autumn 1989): 385–404.

103. Waller, "Funding and Modernization," 392.

104. Rice, *The Hatfields and McCoys,* 50–53.

105. Waller, *Feud,* 5.

106. Ibid., 249.

107. *Pike County News,* 3 Dec. 1931, 23 June 1932.

108. Wells, *Roadside History,* 208, 230, 238.

109. Ibid., 266–68.

110. Waller, *Feud,* 159.

111. *Pike County News,* 1 Oct. 1926, 22 July, 19 Aug. 1937.

112. Helen Powell, "National Register of Historic Places Inventory—Nomination Form, Historic Resources of Pikeville," sec. 7, 7, 11.

113. Dixon, *Chesapeake & Ohio in the Coalfields,* 83, 86.

114. Pike County Historical Society, *Pike County Historical Papers,* 2, 85; Dodrill, *Twenty Thousand Company Stores,* 72, 105.

115. Caudill, *Theirs Be the Power,* 110.

116. Bureau of the Census, *Sixteenth Census, 1940,* 1:425; Lon Rogers, "Fon Rogers: A Pioneer Coal Operator in Pike Company," in Pike County Historical Society, *Pike County, 1822–1976: Historical Papers* 2 (Pikeville, Ky.: Pike County Historical Society, 1976), 85.

117. George Bartley interview, Mar. 1977, AOHP, Alice Lloyd College; Ruby Childers interview, Mar. 1977, AOHP, Alice Lloyd College.

118. Bartley interview.

119. *Appalachian (Pikeville, Ky.) News-Express,* 20 Jan. 1993; Bartley interview.

120. Quoted in Deskins, *Pike County,* 147.

121. *New York Times,* 20 Oct. 1963.

122. *Appalachian (Pikeville, Ky.) News-Express*, 20 Jan. 1993.

123. Rennick, *Place Names of Pike County*, 20.

124. Pike County Historical Society, *Pike County Historical Papers*, 2:92; *Big Sandy (Louisa) News*, 3 Aug. 1917.

125. Dixon, *Chesapeake & Ohio in the Coalfields*, 88.

126. Ibid.

127. *Sandy (Louisa) News*, 2 Feb., 13 Sept. 1912, 15 Oct. 1914; Rennick, *Place Names of Pike County*, 119.

128. Dept. of the Interior, "Historic Resources of Stone, Kentucky," unpublished National Register Survey, 1987.

129. Doris Kearns Goodwin, *The Fitzgeralds and the Kennedys* (New York: Simon and Schuster, 1987), 328–29.

130. *Louisville (Ky.) Courier-Journal*, 8 Nov. 1978. Hale spent most of his life in Stone and was interviewed in 1978; Jim Keenan, "Stone Historical Background Notes," unpublished paper.

131. Allan Nevins and Frank Hill, *Ford: Expansion and Challenge, 1915–1933* (New York: Charles Scribner's Sons, 1957), 645 n; Dept. of the Interior, "Historic Resources," 6.

132. Taylor & Taylor Associates, "Survey Summary Report: Fordson Coal Company Buildings, Stone, Kentucky," 2001 unpublished report, 6, 8.

133. Rennick, *Place Names of Pike County*, 80, 81.

134. Kleber, *Kentucky Encyclopedia*, 546.

135. I. A. Bowles, *A History of Letcher County: Its Political and Economic Growth and Development* (Lexington, Ky.: Hurst Publishing Co., 1949), 37. Eller, *Miners, Millhands, and Mountaineers*, 53–55.

136. Bowles, *History of Letcher County*, 36; Beachley, *History of Consolidation Coal*, 57.

137. Bowles, *History of Letcher County*, 36; Beachley, *History of Consolidation Coal*, 59; Turner and Traum, *John C.C. Mayo*, 37.

138. Turner and Traum, *John C.C. Mayo*, 45.

139. "City of Jenkins was a Farm Site Just Twenty Years Ago," *Pike County (Ky.) News*, 6 Sept. 1928; William Roscoe Thomas, *Life among the Hills and Mountains of Kentucky* (Louisville, Ky.: Standard Printing Co., 1926), 354.

140. *Ashland (Ky.) Daily Independent*, Sept. 5, 1913.

141. Alphonse F. Brosky, "Building a Town for a Mountain Community: A Glimpse of Jenkins and Nearby Villages," *Coal Age* 23 (5 Apr. 1923): 562.

142. Ibid.

143. Beachley, *History of Consolidation Coal*, 91.

144. "Sociological Work Accomplished by the Consolidation Coal Company," *Coal Age* 15 (9 Jan. 1919): 54.

145. Jenkins Heritage Foundation, "The Birth of a Coal Mining Town: Jenkins, Kentucky," videocassette, Jenkins, Ky., 1990.

146. *Ashland (Ky.) Daily Independent,* 5 Sept. 1913.
147. Ibid., 55, 56; Jenkins Heritage Foundation, "Birth of a Coal Mining Town."
148. Beachley, *History of Consolidation Coal,* 71, 77,
149. Federal Writers' Project, *Kentucky,* 437.
150. "The Assimilation of the Foreigner," *Coal Age* 3 (22 Feb. 1913): 307.
151. Jenkins Heritage Foundation, "Birth of a Coal Mining Town."
152. Jenkins received an overall score of 76.5. Some historians have questioned the points awarded southern Appalachian coal towns because they were penalized for not offering regular garbage removal or indoor toilets. These services were common in the north but still rare in southern urban areas in the early 1920s. See Senate, *Report of the U.S. Coal Commission* (Washington, D.C.: GPO, 1925), 3:1435–36.
153. Ibid.
154. *Louisville (Ky.) Courier-Journal,* 11 Dec. 1988.
155. Ibid., 10 Nov. 1991.
156. Dixon, *Chesapeake & Ohio in the Coalfields,* 102.
157. *Lexington (Ky.) Herald-Leader,* 22 June 1987.
158. Rennick, *Kentucky Place Names,* 103.
159. Christine Polly Banks, "Consolidation Coal Company Builds McRoberts," in *Essays in Eastern Kentucky History,* ed. Stuart Sprague (Morehead, Ky.: MSU Foundation, 1984), 6.
160. Banks, "Consolidation Coal Company Builds McRoberts," 6.
161. Much of the information on McRoberts was provided by Mrs. Emma Barker of McRoberts, Kentucky. Some has been taken from Bill Fugate, *McRoberts, Kentucky,* 2 vols. (McRoberts, Ky.: n.p., 1974, 1986).
162. Author interview with Opal Jeanne Tuggle and Alan Tuggle, Seco, Ky., 18 May 1999.
163. Federal Writers' Project, *Kentucky,* 245; Rennick, *Kentucky Place Names,* 316.
164. *Whitesburg (Ky.) Mountain Eagle,* 5 Sept. 1990.
165. Bobby Jean Collins, "Blackey: Its Growth and Decline, 1912–1932," unpublished paper, n.d.
166. Author interview with Judge Ruben Watts, Jenkins, Ky., Aug. 10, 1995; Author interview with Joe Begley, Blackey, Ky., 10 Aug. 1995.
167. Caudill, *Night Comes to the Cumberlands,* 204–5.
168. Ibid.; *Whitesburg (Ky.) Mountain Eagle,* 28 Aug. 1991.
169. Donovan G. Cain, "C. B. Caudill Store," unpublished National Register of Historic Places Registration Form, Whitesburg, Ky., 2001, 5.
170. Studs Terkel, *American Dreams: Lost and Found* (New York: Pantheon Books, 1980), 183, 189; *Coming of Age: The Story of Our Century by Those Who've Lived It* (New York: New Press, 1995).
171. Begley interview.

172. *Lexington (Ky.) Herald-Leader,* 28 Apr. 1999.

173. Zoe Strecker, *Kentucky off the Beaten Path* (Gloster, Conn.: Globe Pequot Press, 1992), 63; William H. Martin, "The Lilley Cornett Woods: A Stable Mixed Mesophytic Forest in Kentucky," *Botanical Gazette* 136 (1975): 171–83; *Lexington Leader,* 4 Dec. 1967.

6. Harlan County and the Big Stone Gap Coalfields

1. Bill Hendrick, *Big Stone Gap: The Early Years* (Big Stone Gap, Va.: n.p., 1990), 9.

2. Wolfe and Wolfe, *Appalachian Coal Hauler,* 6–7.

3. Eller, *Miners, Millhands, and Mountaineers,* 77.

4. Edward R. Tucker, "John R. Fox Jr.," in *Dictionary of Literary Biography: American Novelists, 1910–1945,* ed. James J. Martine (Detroit: Gale Research Co., 1981), 9:25–27.

5. Virginia Historic Landmark Commission, "National Register of Historic Places Inventory—Nomination Form, John Fox Jr. House," sec. 1, 2.

6. Eller, *Miners, Millhands, and Mountaineers,* 59, 60, 76; Luther F. Addington, *The Story of Wise County* (Wise, Va.: Wise County Centennial Committee, 1956), 213–18; Hendrick, *Big Stone Gap,* 5, 6.

7. Wise County Historical Society, *The Heritage of Wise County and the City of Norton, 1856–1993* (Waynesville, N.C.: Professional Press, 1993), 18–19.

8. Ibid., 18–19; Emma Jane James, comp., *As We Were: Memories of Appalachia and Her Surrounding Coal Camps* (Chapel Hill, N.C.: Professional Press, 1993), 104–7; Author interview with Louis Henegar and Bobby Dorton, Appalachia, Va., May 1996.

9. James, *As We Were,* 107.

10. Edwin J. Prescott, *The Story of Virginia Coal and Iron Company, 1882–1945* (Big Stone Gap: Virginia Coal and Iron Co., 1946), 31–52.

11. Ibid., 53–56.

12. Ibid., 61–66.

13. Margaret Ripley Wolfe, "Putting Them in Their Places: Industrial Housing in Southern Appalachia, 1900–1930," *Appalachian Heritage* 7 (summer 1979): 30; Shifflett, *Coal Towns,* 37; Margaret Ripley Wolfe, "Changing the Face of Southern Appalachia: Urban Planning in Southwest Virginia and East Tennessee, 1890–1929," *Journal of Urban Planning* 47 (1981): 252.

14. Shifflett, *Coal Towns,* 54.

15. Prescott, *Virginia Coal and Iron,* 79; Shifflett, *Coal Towns,* 33–35.

16. Prescott, *Virginia Coal and Iron,* 79–82; Shifflett, *Coal Towns,* 37.
17. Shifflett, *Coal Towns,* 64–65.
18. Ibid., 37, 69, 70, 169, 79.
19. Ibid., 54, 37.
20. Ibid., 200, 139–40.
21. Ibid., 209, 211.
22. Prescott, *Virginia Coal and Iron,* illustrations following page 114; Wise County Historical Society, *Heritage of Wise County,* 21.
23. Wise County Historical Society, *The Heritage of Wise County,* 23; Prescott, *Virginia Coal and Iron,* 9: Shifflett, *Coal Towns,* 135.
24. James, *As We Were,* 32; Wolfe and Wolfe, *Appalachian Coal Hauler,* 69–70.
25. Nancy J. Martin-Perdue and Charles L. Perdue Jr., eds., *Talk about Trouble: A New Deal Portrait of Virginians in the Great Depression* (Chapel Hill: Univ. of North Carolina Press, 1996), 329, 335.
26. *Coal Age* 39 (1 Sept. 1934): 367.
27. Shifflett, *Coal Towns,* 64.
28. Ibid., 210.
29. Dixon, *Appalachian Coal Hauler,* 57–58.
30. Dodrill, *Twenty Thousand Company Stores,* 243.
31. Dixon, *Appalachian Coal Hauler,* 58–59.
32. Prescott, *Virginia Coal and Iron,* 102.
33. James, *As We Were,* 40; *Heritage of Wise County,* 24–25.
34. "The Angel of Happy of Happy Hollow," *Saturday Evening Post,* 14 Feb. 1948.
35. Shifflett, *Coal Towns,* 56, 65.
36. *Journal of Urban Planning* 47 (1981): 255; Shifflett, *Coal Towns,* 54.
37. *Journal of Urban Planning* 47 (1981): 255.
38. Prescott, *Virginia Coal and Iron,* illustrations following page 54.
39. Shifflett, *Coal Towns,* 45; Prescott, *Virginia Coal and Iron,* 96.
40. Wise County Historical Society, *Heritage of Wise County,* 26.
41. Kleber, *Kentucky Encyclopedia,* 409.
42. Tony Bubka, "The Harlan County Coal Strike of 1931," *Labor History* 11 (winter 1970): 41–42; John W. Hevener, *Which Side Are You On? The Harlan County Coal Miners, 1931–39* (Urbana: Univ. of Illinois Press, 1978), 3–7; Eller, *Miners, Millhands, and Mountaineers,* 94.
43. Hevener, *Which Side Are You On?* 179–80.
44. Kleber, *Kentucky Encyclopedia,* 409.
45. Dorinda Kim Blackey, "National Register of Historic Places Inventory—Nomination Form, Harlan Commercial District," sec. 6, 3–4.
46. *Lexington (Ky.) Herald-Leader,* 22 Sept. 1992.
47. Mabel Green Condon, *A History of Harlan County* (Nashville, Tenn.: Parthenon Press, 1962), 7–14; Rennick, *Kentucky Place Names,* 131;

Hevener, *Which Side Are You On?* 108, 111; *Louisville (Ky.) Courier-Journal,* 26 Apr. 1991.

48. Dodrill, *Twenty Thousand Company Stores,* 192.

49. Rennick, *Kentucky Place Names,* 96; Bubka, "Harlan County Strike," 47–48.

50. Karen E. Hudson, "National Register of Historic Places Registration Form, Cumberland Central Business District," sec. 7, 3–4.

51. Federal Writers' Project, *Kentucky,* 439; *Harlan (Ky.) Daily Enterprise,* 30 Jan. 1985. Professor Goode has also produced two videotapes dealing with the histories of Benham and Lynch.

52. James B. Goode, *Coal, Steel, Machines, and Men: The Benham Story,* Appalachian Archives, Southeast Community College, Cumberland, Ky., 1990; Rennick, *Kentucky Place Names,* 20.

53. Eller, *Miners, Millhands, and Mountaineers,* 146–47; Goode, *Coal, Steel, Machines, and Men.*

54. W. C. Tucker, "Welfare Work at Benham, Kentucky," *Coal Age* 3 (31 May 1913): 845.

55. Goode, *Coal, Steel, Machines, and Men;* Tucker, "Welfare Work at Benham," 845.

56. Tucker, "Welfare Work at Benham," 846.

57. Goode, *Coal, Steel, Machines, and Men.*

58. Ibid.; H. B. Cooley, "Reconstruction of the Benham Tipple," *Coal Age* 11 (14 Apr. 1920): 662.

59. Hevener, *Which Side Are You On?* 165.

60. Goode, *Coal, Steel, Machines, and Men.*

61. Author interview with director Mrs. Bobbie Gothard, Benham, Ky., May 1996.

62. Philip Thomason, "National Register of Historic Places Inventory—Nomination Form, Benham Historic District," sec. 6, 2–3.

63. Author interview with Jim Whitaker, Benham, Ky., May 1996.

64. Thomas A. Kelemen, "A History of Lynch, Kentucky," *Filson Club History Quarterly* 48 (Apr. 1974): 156; Eller, *Miners, Millhands, and Mountaineers,* 147; Caudill, *Theirs Be the Power,* 98; Rennick, *Kentucky Place Names,* 181.

65. Caudill, *Theirs Be the Power,* 98; Eller, *Miners, Millhands, and Mountaineers,* 172.

66. H. N. Eavenson, "Building Complete Thousand Dwelling Town for a Mine Population of 7,000 at Lynch, Kentucky," *Coal Age* 20 (21 Oct. 1921): 533.

67. Caudill, *Theirs Be the Power,* 95, 98, 99; Eller, *Miners, Millhands, and Mountaineers,* 190.

68. Howard N. Eavenson, "Bathouse, Hospital, and Heating Arrangements Provided for Employees of Lynch Mine in Kentucky," *Coal Age* 20 (27 Oct. 1921): 676; Caudill, *Theirs Be the Power,* 95, 99.

69. Federal Writers' Project, *Kentucky*, 439; Kelemen, "History of Lynch," 167; Author interview with Hugh Webb, Lynch, Ky., May 1996; Thomas R. Ferrall, "News from United States Steel" (n.p., n.d.), 1. Lynch was unincorporated and exact census figures are not available; *Lexington (Ky.) Herald-Leader*, 27 Nov. 1998.

70. Shifflett, *Coal Towns*, 231–32; *Coal Age* 36 (15 Nov. 1931): 606.

71. Hevener, *Which Side Are You On?* 165.

72. Federal Writers' Project, *Kentucky*, 439; Ferrall, "News from United States Steel," 4; Caudill, *Theirs Be the Power*, 102.

73. Kelemen, "History of Lynch," 172; Eavenson, "Building Complete Thousand Dwelling," 536.

74. *New York Times*, 20 Oct. 1963.

75. Site information from author interviews with Tammy Marsili, Hugh Webb, and Michael Lunsford, Lynch, Ky., May 1996.

Bibliography

Articles

Bagger, Eugene S. "Himler of Himlerville." *Survey* 48 (29 Apr. 1922): 146–50.

Bailey, Kenneth R. "A Judicious Mixture: Negroes and Immigrants in the West Virginia Mines, 1880–1917." *West Virginia History* 34 (Jan. 1973): 141–61.

"Bartley Blast Probably Due to Gas Ignition of Electrical Origin, Says Mine Bureau." *Coal Age* 45 (Apr. 1941): 104.

Battlo, Jean. "Mining in the Melting Pot: The African-American Influx into the McDowell County Mines." *Goldenseal* 23 (winter 1997): 46–51.

Bigart, Homer. "Kentucky Miners: A Grim Winter." *New York Times,* 20 Oct. 1963, 1:79.

Blakey, George T. "The New Deal and Rural Kentucky, 1933–1941." *Register of the Kentucky Historical Society* 84 (spring 1986): 146–91.

Brosky, Alphonse F. "Building a Town for a Mountain Community: A Glimpse of Jenkins and Nearby Villages." *Coal Age* 23 (5 Apr. 1923): 560–63.

———. "Sociological Works Accomplished by the Consolidation Coal Company." *Coal Age* 15 (Jan. 1919): 54–58.

Bubka, Tony. "The Harlan County Coal Strike of 1931." *Labor History* 11 (winter 1970) : 41–57.

Cabell, Charles A. "Building a Mining Community." *West Virginia Review* 4 (Apr. 1927): 208–11.

Cantrell, Doug. "Himlerville: Hungarian Cooperative Mining in Kentucky." *Filson Club History Quarterly* 66 (Oct. 1992): 513–42.

Carr, Joe Daniel. "Labor Conflicts in the Eastern Kentucky Coal Fields." *Filson Club History Quarterly* 47 (Apr. 1973): 179–92.

Carter, Alice E. "Segregation and Integration in the Appalachian Coalfields: McDowell County Responds to the *Brown* Decision." *West Virginia History* 54 (Spring 1996): 78–104.

Coleman, J. Winston. "Old Kentucky Iron Furnaces." *Filson Club History Quarterly* 31 (Summer 1957): 227–42.

Cooley, H. B. "Reconstruction of the Benham Tipple." *Coal Age* 11 (14 Apr. 1920): 661–62.

DeVenney, Thomas. "Portsmouth-Solvay Plant in Kentucky." *Coal Age* 12 (24 Nov. 1917): 890–91.

Eavenson, Howard N. "Bathouse Hospital and Heating Arrangement Provided for Employees of Lynch Mine in Kentucky." *Coal Age* 20 (27 Oct. 1921): 676–78.

Egerton, John. "Boom or Bust in the Hollows." *New York Times Magazine*, 18 Oct. 1981.

Fishback, Price V. "Did Coal Miners 'Owe Their Souls to the Company Store?' Theory and Evidence from the Early 1900s." *Journal of Economic History* 46 (Dec. 1986): 1011–29.

"Gas Explosion in Pond Creek Pocahontas Mine Takes a Toll of Ninety-one Lives." *Coal Age* 45 (Feb. 1941): 108.

Hager, Beth A. "Millionaires' Town: The Houses and People of Bramwell." *Goldenseal* 8 (winter 1982): 43–54.

Hamill, R. H. "Design of Buildings in Mining Towns." *Coal Age* 11 (16 June 1917): 1045–48.

Harris, George W. "The Consolidation Coal Company." *Coal Age* 14 (26 Dec. 1918): 1148–53.

Haworth, J. R. "Hungarians Successfully Conduct Co-Operative Mine in Kentucky, Having Two Million Dollars Invested." *Coal Age* 20 (15 Sept. 1921): 412–14.

Hornick, Mike. "Gary Celebrates 100 Years of Mining." *Welch Daily News*, 3 June 1983.

Huebner, A. F. "Houses for Mine Villages." *Coal Age* 12 (27 Oct. 1917): 717–20.

Huddleston, Eugene L. "Russell Terminal in the Days of Steam." *Chesapeake and Ohio Historical Magazine* 33 (Oct. 2001): 3–16.

Hypes, Larry. "No. 14 Had Wild, Woolly Start." *Bluefield Daily Telegraph*, 22 Apr. 1990.

Johnson, James P. "Drafting the NRA Code of Fair Labor Competition for the Bituminous Coal Industry." *Journal of American History* 53 (Dec. 1966): 521–41.

Jordan, Greg. "Wall Comes Tumbling Down at Massive Gary Complex." *Bluefield Daily Telegraph*, 24 June 1991.

Kelemen, Thomas A. "A History of Lynch, Kentucky, 1917–1930." *Filson Club History Quarterly* 48 (Apr. 1974): 156–76.

Loeb, Penny. "Special Report: Shear Madness." *U.S. News and World Report* 123 (11 Aug. 1997): 26–36.

Lohman, K. B. "A New Era for Mining Towns." *Coal Age* 8 (Nov. 1915): 799–800.

Long, Michael E. "Wrestlin' for a Livin' with King Coal." *National Geographic* 163 (June 1983): 792–819.

Maloney, John. "The Angel of Happy Hollow." *Saturday Evening Post*, 14 Feb. 1948, 30.

Martin, William H. "The Lilley Cornett Woods: A Stable Mixed Mesophytic Forest in Kentucky." *Botanical Gazette* 136 (fall 1975): 171–83.

Mathews, Garrett. "Hard Times 'Paralyzing' Gary Hollow." *Bluefield Daily Telegraph*, 26 Sept. 1983.

Matijasic, Thomas D. "The Ku Klux Klan in the Big Sandy Valley of Kentucky." *Journal of Kentucky Studies* 10 (Sept. 1993): 75–80.

McGehee, Stuart. "A Busy Time in McDowell History: Looking Back with John J. Lincoln." *Goldenseal* 15 (fall 1989): 56–64.

———. "Historic Coalwood." *Goldenseal* 27 (summer 2001): 52–59.

Miller, George H. "Plan Your Town as Carefully as You Would Your Plant." *Coal Age* 14 (20 July 1918): 130–31.

Mueller, Lee. "A Look Back at the War the U.S. Brought to Kentucky." *Lexington Herald-Leader*, 23 April 1989.

Munn, Robert F. "The Development of Model Towns in the Bituminous Coal Fields." *West Virginia History* 40 (spring 1979): 243–53.

Osborne, Frank C. "Saint Casimir Mission Center—Where It All Began." *Bankmule: The Official Publication of the Van Lear, Kentucky Historical Society* 3 (Dec. 1986): 4–7.

Peck, W. R., and R. J. Sampson. "The Harlan Coal Field of Kentucky." *Coal Age* 3 (May 24, 1913): 796–800.

Perl, Peter. "How Green Is My Valley Again: When a Coal Town Is Dying." *Washington Post*, 17 Dec. 1984.

Price, E. R. "Wheelwright Program Brings Increased Safety and Efficiency with Lower Operating Costs." *Coal Age* 37 (10 Dec. 1932): 424–25.

Pultz, John Leggett. "The Big Stone Gap Coal-Field of Virginia and Kentucky." *Engineering Magazine* 28 (Oct. 1904): 71–88.

Ratliff, Paul E. "Hellier." *Louisville (Ky.) Courier-Journal,* 18 Dec. 1977, 46.

Savage, Joe E. "The Armed March in West Virginia." *Goldenseal* 20 (winter 1994): 54–60.

Semrau, Ronda G. "Roxie Gore: Looking Back in Logan County." *Goldenseal* 16 (summer 1990): 23–28.

Shumway, R. W. "Belt Conveyor Transfers Coal across River with Suspension Bridge as Support." *Coal Age* 19 (24 Mar. 1921): 525–27.

Simon, Richard M. "Uneven Development and the Case of West Virginia: Going Beyond the Colonialism Model." *Appalachian Journal* 8 (spring 1981): 165–86.

Tucker, Edward R. "John R. Fox Jr." In *Dictionary of Literary Biography: American Novelists, 1910–1945,* edited by James J. Martine, 9:25–27 Detroit: Gale Research Co., 1981.

Tucker, William C. "Welfare Work at Benham, Kentucky." *Coal Age* 3 (31 May 1913): 844–46.

Waller, Altina L. "Feuding and Modernization in Appalachia: The Hatfield and McCoys." *Register of the Kentucky Historical Society* 87 (autumn 1989): 385–404.

Webb, William B. "The Elkhorn Coal Co.'s Plant." *Coal Age* 10 (12 Aug. 1916): 264–65.

Westleigh, Peter. "Safety the First Consideration." *Coal Age* 3 (1 Mar. 1913): 347–49.

Wolfe, Margaret Ripley. "Changing the Face of Southern Appalachia: Urban Planning in Southwest Virginia and East Tennessee, 1890–1929." *Journal of American Planning* 47 (July 1981): 252–65.

———."Putting Them in Their Place: Industrial Housing in Southern Appalachia, 1900–1930." *Appalachian Heritage* 7 (summer 1979): 27–36.

Unpublished Sources

Agsten, Carl F. "Chafin House." National Register of Historic Places Registration Form. Charleston, W.Va., 1994.

Blackey, Dorinda Kim M. "Harlan Commercial District." National Register of Historic Places Inventory—Nomination Form. Lexington, Ky., 1985.

Bond, John W. "Matewan Historic District." National Historic Landmark Nomination Form. Cherry Hill, N.J., 1996.

Cain, Donovan G. "C. B. Caudill Store." National Register of Historic Places Inventory—Nomination Form. Whitesburg, Ky., 2001.

Collins, Bobby Jean. "Blackey: Its Growth and Decline, 1912–1932." Whitesburg, Ky., n.d.

Edwards, David R. "Pocahontas Mine No. 1." National Historic Landmark Nomination Form. Richmond, Va., 1993.

Gillenwater, Mack H. "Cultural and Historical Geography of Mining Settlements in the Pocahontas Coal Fields of Southern West Virginia, 1880–1930." Ph.D. diss., University of Tennessee, 1972.

Gioulis, Michael. "Bramwell Additions Historic District." National Register of Historic Places Registration Form. Sutton, W.Va., 1995.

Gioulis, Michael, Paul McAllister, and Stacy Sone. "Matewan Historic District." National Register of Historic Places Registration Form. Charleston, W.Va., 1993.

Hager, Beth. "Bramwell Historic District." National Register of Historic Places Inventory—Nomination Form. Charleston, W.Va., 1982.

"Historic Resources of Stone, Kentucky." Unpublished paper prepared for the National Register of Historic Places. Frankfort, Ky., 1988.

Hudson, Karen E. "Cumberland Central Business District." National Register of Historic Places Nomination Form. Lexington, Ky., 1995.

Jones, Calvin, and Gloria Mills. "Wheelwright Commercial District." National Register of Historic Places Inventory—Nomination Form. Frankfort, Ky., 1980.

Keenan, Jim. "Stone Historical Background Notes." Unpublished report, Stone, Ky., 2000.

McGehee, Stuart. "Gary: A First-Class Mining Operation." Unpublished paper, Eastern Regional Coal Archives.

Perry, L. Martin. "Coal Company Towns in Eastern Kentucky, 1854–1941." Unpublished Report of the Kentucky Heritage Council, Frankfort, Ky., 1991.

Pike County Kentucky Fiscal Court. "Regeneration of Stone, KY." Unpublished Report of the Pike County Fiscal Court, Pikeville, Ky., 2000.

Powell, Helen. "Historic Resources of Pikeville." National Register of Historic Places Inventory—Nomination Form. Richmond, Ky., 1983.

Semonco, Joan. "Elkhorn, West Virginia 24831." Unpublished paper, n.d.

Sone, Stacy. "Carter Coal Company Store." National Register of Historic Places Registration Form. Charleston, W.Va., 1992.

———. "Coal Company Stores in McDowell County." National Register of Historic Places Multiple Property Documentation Form. Charleston, W.Va., 1992.

————. "John J. Lincoln House." Unpublished National Register of Historic Places Registration Form. Charleston, W.Va., 1992.

Stallard, Mae. "A Walking Tour of Williamson." Unpublished paper of the Tug Valley Chamber of Commerce, Williamson, W.Va., n.d.

Taylor & Taylor Associates. "Survey Summary Report: Fordson Coal Company Buildings, Stone, Pike County, Kentucky." Unpublished report prepared for Stone Heritage and the Kentucky Heritage Council, Brookville, Pa., 2001.

Thomason, Philip. "Benham Historic District." National Register of Historic Places Inventory—Nomination Form. Frankfort, Ky., 1983.

Turley, Charles E. "McDowell County Courthouse." National Register of Historic Places Inventory—Nomination Form. Charleston, W.Va., 1979.

Virginia Historic Landmarks Commission Staff. "John R. Fox Jr., House." National Register of Historic Places Inventory—Nomination Form. Richmond, Va., 1973.

Valente, Kim A. "Welch Commercial Historic District." National Register of Historic Places Registration Form. Charleston, W.Va., 1991.

Zegeer, David. "Mechanization of the Coal Industry in Appalachia." Unpublished paper, 1999. Author's collection.

Books

Addington, Luther F. *The Story of Wise County, Virginia.* Wise, Va.: Wise County Centennial Committee, 1956.

Baker, Nancy Virginia. *Bountiful and Beautiful: A Bicentennial History of Buchanan County, Virginia, 1776–1976.* Grundy, Va.: Buchanan Vocational School, 1976.

Barnum, Donald T. *The Negro in the Bituminous Coal Industry.* Philadelphia: University of Pennsylvania Press, 1970.

Beachley, Charles E. *History of the Consolidation Coal Company, 1864–1934.* New York: Consolidation Coal Co., 1934.

Berry, Wendell. *Civilizing the Cumberlands.* Lexington, Ky.: King Library Press, 1927.

Bowles, I. A. *History of Letcher County, Kentucky: Its Political and Economic Growth and Development.* Lexington, Ky.: Hurst Publishing Co., 1949.

Bowman, Mary Jean and W. Warren Haynes. *Resources and People in East Kentucky: Problems and Potentials of a Lagging Economy.* Baltimore, Md.: Johns Hopkins University Press, 1963.

Brown, Harlan R. *In the Foothills of the Cumberlands: A History of Eastern Kentucky.* Ashland, Ky.: Graber Printing Co., 1959.

Brown, Sharon A. *A Historic Resource Study: Kaymoor, New River Gorge National Park, West Virginia.* Washington, D.C.: U.S. Dept. of the Interior, 1990.

Caudill, Harry Monroe. *Night Comes to the Cumberlands: A Biography of a Depressed Area.* Boston: Little, Brown & Co., 1963.

———. *Theirs Be the Power: The Moguls of Eastern Kentucky.* Urbana: University of Illinois Press, 1983.

Channing, Steven A. *Kentucky: A History.* New York: W.W. Norton & Co., 1977.

Clark, Thomas D. *A History of Kentucky.* 1937. Reprint, Ashland, Ky.: Jesse Stuart Foundation, 1992.

Cohen, Stan. *King Coal: A Pictorial History of West Virginia Coal Mining.* Charleston, W.Va.: Pictorial Histories Publishing Co., 1984.

Condon, Mabel Green. *A History of Harlan County.* Nashville, Tenn.: Parthenon Press, 1962.

Conley, Phil. *History of the West Virginia Coal Industry.* Charleston, W.Va.: Education Foundation, 1960.

Corbin, David A. *Life, Work, and Rebellion in the Coalfields: The Southern West Virginia Miners, 1880–1922.* Urbana: University of Illinois Press, 1981.

Cox, Clayton R., comp. *Appalachia Crossroads: Descendants of Hezekiah Sellards.* Baltimore, Md.: Gateway Press, 1977.

Crawford, Margaret. *Building the Workingman's Paradise: The Design of American Company Towns.* New York: Verso, 1995.

Crowe-Carraco, Carol. *The Big Sandy.* Lexington: University Press of Kentucky, 1979.

DeRosett, Lou. *Middlesborough: The First Century.* Jacksboro, Tenn.: Action Printing, 1988.

DeRosett, Lou, and Joe Marcum, eds. *Middlesborough: At One Hundred, 1890–1990.* Middlesboro, Ky.: Centennial Commission, 1990.

Deskins, William David. *Pike County: A Very Different Place.* Pikeville, Ky.: Printing by George, 1994.

Dix, Keith. *Work Relations in the Coal Industry: The Hand Loading Era, 1880–1930.* Morgantown, W.Va.: Institute for Labor Studies, 1977.

———. *What's a Coal Miner to Do? The Mechanization of Coal Mining.* Pittsburgh: University of Pittsburgh Press, 1988.

Dixon, Thomas W. *Appalachian Coal Mines and Railroads.* Lynchburg, Va.: TLC Publishing, 1994.

———. *Chesapeake & Ohio in the Coal Fields.* Clifton Forge, Va.: Chesapeake & Ohio Historical Society, 1995.

Dodrill, Gordon, comp. *Twenty Thousand Coal Company Stores in the United States, Mexico, and Canada.* Pittsburgh: University of Pittsburgh Press, 1971.

Dubofsky, Melvyn, and Warren Van Tine. *John L. Lewis: A Biography.* Urbana: University of Illinois Press, 1986.

Eller, Ronald D. *Miners, Millhands, and Mountaineers: Industrialization of the Appalachian South, 1880–1930.* Knoxville: University of Tennessee Press, 1982.

Ely, William. *The Big Sandy Valley: A History of the People and Country from the Earliest Settlement to the Present Time.* 1887. Reprint, Cattlettsburg, Ky.: Genealogical Publishing Co., 1969.

Faragher, John Mack. *Daniel Boone: The Life and Legend of an American Pioneer.* New York: Henry Holt and Co., 1992.

Federal Writers' Project. *Kentucky: A Guide to the Bluegrass State.* New York: Harcourt, Brace, and Co., 1939.

Fink, Gary L., and Merl E. Reed, eds. *Essays in Southern Labor History: Selected Papers, Southern Labor History Conference, 1976.* Westport, Conn.: Greenwood Press, 1977.

Fischer, David Hackett. *Albion's Seed.* New York: Oxford University Press, 1991.

Ford, Thomas R. *The Southern Appalachian Region: A Survey.* Lexington: University Press of Kentucky, 1967.

Francavaglia, Richard V. *Hard Places: Reading the Landscape of America's Historic Mining Districts.* Iowa City: University of Iowa Press, 1991.

Fugate, Bill. *McRobert, Kentucky.* 2 vols. McRoberts, Ky.: n.p., 1974, 1986.

Fuson, Henry Harvey. *History of Bell County.* 2 vols. New York: Hobson Book Press, 1947.

Garvey, Thomas H. *The Last Steam Railroad in America.* New York: Harry N. Abrams, 1995.

Geiger, Joe, ed. *Marking Our Past: West Virginia's Historical Highway Markers.* Charleston: West Virginia Division of Culture and History, 2002.

Goodrich, Carter. *The Miner's Freedom: A Study of the Working Life in a Changing Industry.* New York: Arno Press, 1977.

Goodwin, Doris Kearns. *The Fitzgeralds and the Kennedys.* New York: Simon and Schuster, 1987.

Hall, C. Mitchell. *Jenny Wiley Country: A History of the Big Sandy Valley in Kentucky's Eastern Highlands and Genealogy of the Region's People.* 2 vols. Kingsport, Tenn.: Professional Press, 1972–79.

————. *Johnson County, Kentucky: A History of the County and Genealogy of Its People up to the Year 1927*. 2 vols. Louisville, Ky.: Standard Press, 1928.

Hall, R. Dawson, and J. H. Edwards. *Coal Age Mining Manual*. New York: Coal Age, 1934.

Harrison, Lowell H., ed. *Kentucky's Governors, 1792–1985*. Lexington: University Press of Kentucky, 1985.

Hendrick, Bill. *Big Stone Gap: "The Early Years."* Norton, Va.: n.p., 1990.

Hevener, John W. *Which Side Are You On ? The Harlan County Coal Miners, 1931–39*. Urbana: University of Illinois Press, 1978.

Hood, Fred J., ed. *Kentucky: Its History and Heritage*. St. Louis, Mo.: Forum Press, 1978.

Jakle, John H., Robert W. Bastian, and Douglas K. Meyer. *Common Houses in America's Small Towns*. Athens: University of Georgia Press, 1989.

James, Emma Jane. *As We Were: Memories of Appalachia and Her Surrounding Coal Camps*. Chapel Hill, N.C.: Professional Press, 1993.

Johnson County Historical and Genealogical Society. *Historic Sites of Paintsville and Johnson County, Kentucky*. Frankfort, Ky.: Kentucky Heritage Council, 1985.

Jones, G. C. *Growing Up Hard in Harlan County*. Lexington: University Press of Kentucky, 1985.

Jones, Jack M. *Early Mining in Pocahontas, Virginia*. Lynchburg, Va.: Jack M. Jones, 1983.

Kleber, John E., ed. *The Kentucky Encyclopedia*. Lexington: University Press of Kentucky, 1992.

King, Ed. *Norfolk & Western in the Appalachians: From the Blue Ridge to the Big Sandy*. Waukesha, Wis.: Kalmbach Books, 1998.

Lambie, Joseph T. *From Mine to Market: The History of Coal Transportation on the Norfolk & Western Railway*. New York: New York University Press, 1954.

Lane, Winthrop D. *Civil War in West Virginia: A Story of the Industrial Conflict in the Coal Mines*. New York: B. W. Huebsch, 1921.

Lee, Howard B. *Bloodletting in Appalachia: A Story of West Virginia's Four Major Mine Wars and Other Thrilling Incidents of Its Coal Fields*. Morgantown: West Virginia University Press, 1969.

Long, Priscilla. *Where the Sun Never Shines: A History of America's Bloody Coal Industry*. New York: Paragon House, 1989.

Longstreth, Richard. *The Building of Main Street: A Guide to American Commercial Architecture*. Washington, D.C.: Preservation Press, 1987.

Lunt, Richard D. *Law and Order vs. the Miners: West Virginia, 1907–1933*. Hamden, Conn.: Archon Books, 1979.

Martin-Perdue, Nancy J., and Charles L. Perdue Jr., eds. *Talk about Trouble: A New Deal Portrait of Virginians in the Great Depression*. Chapel Hill: University of North Carolina Press, 1996.

McGehee, Stuart, and Eva McGuire. *A Century of Stewardship: The History of Pocahontas Land Corporation*. Bluefield, W.Va.: Pocahontas Land Corp., 2001.

Miller, Donald L., and Richard E. Sharpless. *The Kingdom of Coal: Work, Enterprise, and Ethnic Communities in the Mine Fields*. Philadelphia: University of Pennsylvania Press, 1985.

Nelson, James Poyntz. *The History of the Chesapeake & Ohio Railway Company: Its Antecedents and Subsidiaries*. Richmond, Va.: Lewis Printing Co., 1927.

Nevins, Allan, and Frank Ernest Hill. *Ford: Expansion and Challenge, 1915–1933*. New York: Charles Scribner's Sons, 1957.

North, E. Lee. *The 55 West Virginias*. Morgantown: West Virginia University Press, 1985.

Parkinson, George. *Guide to Coal Mining Collections in the United States*. Morgantown: West Virginia University Press, 1978.

Perry, Robert. *The Oldest House in the Valley: A Study of the May House in Prestonsburg, Kentucky and the Man Who Built It*. Prestonsburg, Ky.: Floyd Press, 1993.

Peskins, Allan. *Garfield: A Biography*. Kent, Ohio: Kent State University Press, 1978.

Pike County Historical Society. *150 Years of Pike County, Kentucky, 1822–1972*. Vol. 1, sesquicentennial issue. Lexington, Ky.: Transylvania Printing Co., 1972.

———. *Pike County, Kentucky, 1822–1976*. Historical Papers, No. 2. Pikeville, Ky.: Pike County Historical Society, 1976.

Prescott, Edward J., comp. *The Story of the Virginia Coal and Iron Company*. Big Stone Gap: Virginia Coal and Iron Co., 1946.

Preston, John David. *The Civil War in the Big Sandy Valley of Kentucky*. Baltimore, Md.: Gateway Press, 1984.

Raitz, Karl B., and Richard Ulack. *Appalachia: A Regional Geography*. Boulder, Co.: Westview Press, 1984.

Rand-McNally. *2002 Commercial Atlas and Marketing Guide*. 131st ed. Chicago, Ill.: Rand-McNally, 2002.

Rennick, Robert M. *Kentucky Place Names.* Lexington: University Press of Kentucky, 1984.

———. *Place Names of Pike County, Kentucky.* Lake Grove, Ore.: Depot, 1991.

Rice, Otis K. *The Allegheny Frontier: West Virginia Beginnings, 1730–1830.* Lexington: University Press of Kentucky, 1970.

———. *The Hatfields and the McCoys.* Lexington: University Press of Kentucky, 1982.

———. *West Virginia: A History.* Lexington: University Press of Kentucky, 1985.

Rist, Donald E. *Kentucky Iron Furnaces of the Hanging Rock Iron Region.* Ashland, Ky.: Hanging Rock Press, 1974.

Robinson, George W., ed. *Bert Combs, the Politician: An Oral History.* Lexington: University Press of Kentucky, 1991.

Ross, Charlotte, ed. *Bibliography of Southern Appalachia.* Boone, N.C.: Appalachian Consortium Press, 1976.

Ross, Malcolm. *Machine Age in the Hills.* New York: Macmillan, 1933.

Salvati, Raymond E. *Island Creek: A Career Company Dedicated to Coal, Saga in Bituminous.* New York: Newcom Society of North America, 1957.

Savage, Lon K. *Thunder in the Mountains: The West Virginia Mine War, 1920–1921.* Pittsburgh: University of Pittsburgh Press, 1990.

Scalf, Henry P. *Floyd County Sesquicentennial: 1800–1950.* Prestonsburg, Ky.: Floyd Press, 1950.

———. *Kentucky's Last Frontier.* 2d ed. Pikeville, Ky.: Pikeville College Press, 1972.

Schweitzer, Robert, and Michael W. R. Davis. *America's Favorite Homes: Mail-Order Catalogues as a Guide to Popular Early Twentieth Century Houses.* Detroit: Wayne State University Press, 1990.

Seltzer, Curtis. *Fire in the Hole: Miners and Management in the American Coal Industry.* Lexington: University Press of Kentucky, 1985.

Shackleford, Laurel, and Bill Weinberg. *Our Appalachia: An Oral History.* New York: Hill and Wang, 1977.

Shapiro, Henry David. *Appalachia on Our Mind: The Southern Mountains and Mountaineers in the American Consciousness.* Chapel Hill: University of North Carolina Press, 1978.

Shifflett, Crandall A. *Coal Towns: Life, Work, and Culture in Company Towns of Southern Appalachia, 1880–1930.* Knoxville: University of Tennessee Press, 1991.

Sprague, Stuart, ed. *Essays in Eastern Kentucky History.* Morehead, Ky.: MSU Foundation, 1984.

Stilgoe, John R. *Common Landscapes of America, 1580 to 1845.* New Haven, Conn.: Yale University Press, 1982.

———. *Metropolitan Corridor: Railroads and the American Scene.* New Haven, Conn.: Yale Univ. Press, 1982.

Strecker, Zoe. *Kentucky off the Beaten Path.* Gloster, Conn.: Globe Pequot Press, 1992.

Sulzer, Elmer G. *Ghost Railroads of Kentucky.* Indianapolis: Vane A. Jones Co., 1967.

Tackett, Regina, Patricia Jackson, and Janice Thompson, eds. *History of Lawrence County, Kentucky.* Dallas: Curtis Media Corp., 1991.

Tams, W. P., Jr. *The Smokeless Coalfields of West Virginia.* Morgantown: West Virginia University Press, 1983.

Tate, Charles. *Cable Television in the Cities.* Washington, D.C.: Urban Institute, 1971.

Terkel, Studs. *American Dreams: Lost and Found.* New York: Pantheon Books, 1980.

———. *Coming of Age: The Story of Our Century by Those Who've Lived It.* New York: New Press, 1995.

Thomas, William Roscoe. *Life among the Hills and Mountains of Kentucky.* Louisville, Ky.: Standard Printing Co., 1926.

Thurmond, Walter R. *The Logan Coal Field of West Virginia: A Brief History.* Morgantown: West Virginia University Library, 1964.

Trotter, Joe William, Jr. *Coal, Class, and Color: Blacks in Southern West Virginia, 1915–1932.* Urbana: University of Illinois Press, 1990.

Turner, Carolyn Clay, and Carolyn Hay Traum. *John C.C. Mayo: Cumberland Capitalist.* Pikeville, Ky.: Pikeville College Press, 1983.

Turner, William H., and Edward J. Cabbell, eds. *Blacks in Appalachia.* Lexington: University Press of Kentucky, 1985.

Vaughan, James. *Blue Moon over Kentucky: A Biography of Kentucky's Troubled Highlands.* N.p., 1985.

Verhoeff, Mary. *The Kentucky Mountains: Transportation and Commerce, 1750–1911: A Study in the Economic History of a Coal Field.* Filson Club Publication No. 26, vol. 1. Louisville, Ky.: Filson Club, 1911.

Virginia Writers' Program. *Virginia: A Guide to the Old Dominion.* New York: Oxford University Press, 1940.

Waller, Altina L. *Feud: Hatfields, McCoys, and Social Change in Appalachia, 1860–1900.* Chapel Hill: University of North Carolina Press, 1988.

Walters, James E. *Balkan: The Land and People.* Pineville, Ky.: Eighteenth Annual Homecoming Celebration Committee, 1986.

Way, William, Jr. *The Clinchfield Railroad: The Story of a Trade Route across the Blue Ridge Mountains.* Chapel Hill: University of North Carolina Press, 1931.

Webb, William S. *The C & O Mounds.* Lexington: University Press of Kentucky, 1942.

Weller, Jack. *Yesterday's People: Life in Contemporary Appalachia.* Lexington: University Press of Kentucky, 1965.

Wells, Dianne, comp. *Roadside History: A Guide to Kentucky Highway Markers.* Frankfort: Kentucky Historical Society, 2002.

West Virginia Writers' Program. *West Virginia: A Guide to the Mountain State.* New York: Oxford University Press, 1941.

Williams, John Alexander. *West Virginia: A History.* New York: Norton Co., 1984.

———. *West Virginia and the Captains of Industry.* Morgantown: West Virginia University Library, 1976.

Williams, Lewis M. *The Transformation of a Coal Mining Town.* Chicago: Inland Steel Co., n.d.

Wise County Historical Society. *The Heritage of Wise County and the City of Norton, 1856–1993.* Waynesville, N.C.: Professional Press, 1993.

Wolfe, Hugh, and Ed Wolfe. *Appalachian Coal Hauler: The Interstate Railroad's Mine Runs and Coal Trains.* Lynchburg, Va.: TLC Publishing, 2001.

Wolfford, George. *Lawrence County: A Pictorial History.* Ashland, Ky.: W.W.W. Co., 1972.

Government Documents

Rockefeller, John D. *The American Coal Miner: A Report on Community and Living Conditions in the Coalfields.* Washington, D.C.: GPO, 1980.

U.S. Bureau of the Census. *Ninth Census of the United States: 1870.* Washington, D.C.: GPO, 1872.

———. *Fourteenth Census of the United States: 1920.* Washington, D.C.: GPO, 1921.

———. *Eighteenth Decennial Census of the United States: 1960.* Washington, D.C.: GPO, 1962.

U.S. Congress. Senate. *Report of the U.S. Coal Commission.* 68th Cong., 2d sess., 1925. S. Doc. 195.

U.S. Dept. of the Interior, National Parks Service. *Coal Fields, Communities, and Change: Status of Planning, Study of Coal Mining Heritage and Related Resources in Southern West Virginia.* Philadelphia: National Parks Service Mid-Atlantic Regional Office, 1991.

———. *A Coal Heritage Study: A Study of Coal Mining and Related Resources in Southern West Virginia.* Philadelphia: National Parks Service, Mid-Atlantic Regional Office, 1993.

White, Joseph H. *Houses for Mining Towns.* Washington, D.C.: U.S. Dept. of the Interior, 1914.

Newspapers

Kentucky

Ashland Daily Independent, 1900–1920
Big Sandy (Louisa) News, 1900–1920
Floyd County Times, 1970–96
Letcher County Mountain Eagle, 1970–96
Lexington Herald-Leader, 1912–96
Paintsville Herald, 1920–90
Pike County News, 1922–96

West Virginia

Bluefield Daily Telegraph, misc. issues, 1990–98.
Welch Daily News, misc. issues, 1980–96.
Williamson Daily News, 1970–96

Oral Histories

Appalachian Oral History Project (AOHP), Alice Lloyd College, Pippa Passes, Ky.

Bartley, George. Interview by Aloma Faye Ratliff. Hellier, Ky., 14 Mar. 1977.

Bentley, Liva. Interview by Joey Elswick. Stone Coal, Ky., 20 June 1976.

Childers, Ruby. Interview by Aloma Faye Ratliff. Hellier, Ky., 30 Mar. 1977.

Garrett, Hilton. Interview by Luther Frazier. Wheelwright, Ky., 8 Aug. 1973.

Osborne, Dewey. Interview by Patti Rose. Wheelwright, Ky., 29 June 1971.

Spence, Opal Wooten. Interview by Aloma Faye Ratliff. Hellier, Ky., 16 April 1977.

Spencer, Rudolph. Interview by Aloma Faye Ratliff. Garrett, Ky., 8 Sept. 1982.

Interviews with Authors

Begley, Joseph. Blackey, Ky., 10 Aug. 1995.

Dorton, Bobby. Appalachia, Va., 22 May 1996.

Gothard, Bobbie. Benham, Ky., 25 May 1996.

Henegar, Louis. Appalachia, Va., 22 May 1996.

Lunsford, Michael. Lynch, Ky., 25 May 1996.

Marsili, Tammy. Lynch, Ky., 25 May 1996.

Owens, John. Pikeville, Ky., 10 Sept. 1992.

Reynolds, David. Matewan, W.Va., 28 May 1996.

Watts, Ruben. Jenkins, Ky., 10 Aug. 1995.

Webb, Hugh. Lynch, Ky., 25 May 1996.

Whitaker, James. Benham, Ky., 30 May 1996.

Wright, Raymond. McDowell, Ky., 20 Dec. 1995.

Zegeer, David. Lexington, Ky., 10 June 1999.

Video Productions

Goode, James B. *Coal, Steel, Machines, & Men: The Benham Story.* Appalachian Archives, Cumberland, Ky., 1990.

Jenkins Heritage Foundation. *The Birth of a Coal Mining Town: Jenkins, Kentucky.* Jenkins Heritage Foundation, Jenkins, Ky., 1990.

Web Sites

http://bramwellwv.com

http://www.coaleducation.org

http://www.cohs.org

http://www.frograil.com

http://imagebase.lib.vt.edu/search.php

http://www.jsfbooks.com

http://www.matewan.com

http://nationalregisterofhistoricplaces.com

http://www.northeast.railfan.net

http://www.nr.nps.gov

http://www.msha.gov

http://www.spikesys.com

http://www.visitashlandky.com

http://www.williamsonrailroadmuseum.com

http://www.wvculture.org/history/journal_wvh

Miscellaneous Sources

Boone County Community and Economic Development Office, Danville, W.Va.

Bramwell Millionaire Garden Club, Bramwell, W.Va.

Chesapeake & Ohio Historical Society, Cumberland Tourism Commission, Cumberland, Ky.

Kentucky Coal Mining Museum, Benham, Ky.

Mercer County Convention and Visitors Bureau, Bluefield, W.Va.

Pikeville–Pike County Tourism Commission, Pikeville, Ky.

West Virginia Coal Heritage Trail Association, Bluefield, W.Va.

Wise County, Virginia, Chamber of Commerce.

Index

Peach Orchard Coal Company, 171
Perkins, Carl D., 54
Pike County, Ky., 16, 22, 172, 209–29
Pikeville College, 214
Pikeville, Ky., 13, 25, 32, 77, 99, 104,
    192, 210–16
Pine Branch, Va., 94, 280
Pine Branch coke ovens, 280–81
Pine Mountain, 251
PMWA. See Progressive Mine Workers
    of America.
Progressive Mine Workers of America
    (PMWA), 289–90, 296
Pocahontas Coal and Coke
    Company, 149
Pocahontas Exhibition Mine, 157
Pocahontas Land Company, 127
Pocahontas, Va., 16, 36, 91, 96, 107, 144,
    154–58
Pond Creek, 223, 224
Pond Creek Coal Company, 125,
    224, 225
Pond Creek Pocahontas Coal Company,
    140, 141
Powell River, 258
Powerhouses, 95
Preston, John, 192
Prestonsburg, Ky., 30, 32, 56, 190,
    192–94
Price, Emory R. (Jack), 13, 203, 208
Price, Ky., 208
Princess Furnace, 166
Privies, 97–98
Public Works Administration (PWA), 87

Railroad Stations, 103
Railroad Yards, 102
Railroads, 14–18, 100
Red Jacket Consolidated Coal and Coke
    Company, 124, 125
Red Jacket, W.Va., 124–26
Regional Commercial Centers, 76–77
Religion, 85
Richardson, George S., 172
Richardson, Ky., 171–74. See also Peach
    Orchard, Ky.
River, Ky., 98
Roda, Va., 267, 269, 273–75

Room and pillar method, 62
Rush, Ky., 163
Russell Fork, 2, 3, 222, 230
Russell, Ky., 15, 102, 164–65

SECO. See Southeast Coal Company.
Sayles, John, 107, 121
Schools, 86–87
Scrip, 88
Seco, Ky., 25, 247–50
Shelbiana, Ky., 87
Shelby Yard, 217
Sixteen Tons, 235
Sixth Street Furnace, 166
Southwest Virginia Improvement
    Company, 127
Southwest Virginia Museum, 261
Spencer, Ann, 153
Steel, 60
Stone buildings, 87, 106
Stone, Edward, 109
Stone, Galen L., 224, 225
Stone, Ky., 87, 111, 224–29
Stonega Coke and Coal Company, 18,
    37, 85, 257, 258, 262, 264–65, 267,
    271, 273
Stonega, Va., 16, 73, 76, 265–69, 275–78
Strip mining, 9, 55, 90
Stuart, Jesse, 167–68
Surface Mining Control and
    Reclamation Act (1977), 254

Taggart, John, 257, 265–66
Terkel, Studs, 254–55
Testerman, Cabell C., 42, 123, 124
Tipples, 91–93
Thacker, W. Va., 124
Thomas, William, 150–51
Trail of the Lonesome Pine, 259–60
Tug Fork, 2, 13, 16, 22, 36, 107, 112,
    128, 134, 210
Turner-Elkhorn Mining Company, 198

United Mine Workers of America
    (UMWA), 39, 40, 50, 51, 125–26,
    133, 137, 142, 195, 200, 202, 247,
    268, 282, 283, 289–90, 296
U.S. 23, 19, 20–22

A Guide to Historic Coal Towns of the Big Sandy River Valley was designed and typeset on a Macintosh computer system using QuarkXPress software. The body text is set in 10.5/14 Minion with display type set in Helvetica Neue and Industria. This book was designed and typeset by Cheryl Carrington and manufactured by Thomson-Shore, Inc.

www.ingramcontent.com/pod-product-compliance
Lightning Source LLC
Chambersburg PA
CBHW021847020426
42334CB00013B/217